# Managing Food Insecurity Risk

ANALYTICAL FRAMEWORK
AND APPLICATION TO INDONESIA

**Please cite this publication as:**
OECD (2015), *Managing Food Insecurity Risk: Analytical Framework and Application to Indonesia*, OECD Publishing, Paris.
*http://dx.doi.org/10.1787/9789264233874-en*

ISBN 978-92-64-23386-7 (print)
ISBN 978-92-64-23387-4 (PDF)



**Photo credits:** Cover © joloei/iStock/Thinkstock

Corrigenda to OECD publications may be found on line at: *www.oecd.org/about/publishing/corrigenda.htm*.

# *Foreword*

Many of the recent concerns about food security relate to perceived threats to current levels of food security, such as those due to price shocks or natural disasters. These threats concern the stability dimension of food security, increasing so-called transitory food insecurity. The present publication describes an approach designed to explore effective policy responses to this risk environment within a broader policy strategy for improving the long-term availability, access and utilisation of food.

The publication describes and illustrates a consolidated approach for guiding policy choices aiming to deal with transitory food insecurity. Called the Framework for Analysing Transitory Food Insecurity, it is a tool that first assesses the risks of food insecurity in a given context, and then, using a risk management approach, examines different policy instruments available to government for addressing the main food insecurity threats identified. It is designed to be a tool for examining the robustness of policy responses to managing risks and uncertainty across a variety of different threats to food security.

After a conceptual and theoretical overview, this framework is then applied to a case study for Indonesia. The analysis reported is the result of very intensive exchanges with Indonesian policy makers, experts and stakeholders. The case study followed the guidelines defined in the framework, in particular the three-step process comprising Preparatory Work, Risk Assessment and Policy Dialogue. Three OECD missions to Indonesia were associated with these steps: a preparatory mission in June 2013, a risk assessment mission in October 2013, and a policy consultation mission in February 2014. A regional conference on policies for food security held in Bogor in November 2014 provided an opportunity for further discussion of the policy implications of the study.

This publication synthesises the work related to the management of food insecurity risk in six sections. Section 1 gives an overview of the concepts, measurement issues and causes associated with food insecurity risk, while Section 2 presents the theoretical basis for the analytical approach adopted. Sections 3, 4 and 5 report the procedures used, and the results obtained from applying each of the three steps of the framework to the situation in Indonesia. Section 6 draws on the results of the three steps to provide further policy discussion, conclusions and recommendations

The reports on which this publication is based were declassified by the Working Party for Agricultural Policy and Markets of the OECD Committee for Agriculture in November 2014.

## *Acknowledgements*

This publication was prepared by the Trade and Agriculture Directorate of the OECD with the active participation of Indonesian authorities and experts. Jesus Antón and Shingo Kimura of the OECD contributed to the coordination and drafting of this report. Statistical assistance was provided by Joanna Ilicic-Komorowska and Kentaro Numa. Administrative and editing services were provided by Michèle Patterson, Martina Abderrahmane, Marina Giacalone-Belkadi and Alison Burrell. Andrzej Kwiecinski provided useful comments on the Indonesian case study. This work has also benefitted from the discussions at OECD Working Party meetings, academic conferences, workshops, seminars, and informal meetings with policy makers, stakeholders, industry representatives, experts and other colleagues.

The Indonesian case study was conducted in close collaboration with the Indonesian Ministry of Agriculture, with particular support received from Tahlim Sudaryanto (Assistant Minister for International Cooperation). Two roundtables involving stakeholders and experts were organised in Indonesia to identify the perceived risks and threats to food security. A high-level seminar was also organised to discuss the results of risk assessment of food insecurity scenarios and policy impacts. This seminar was attended by senior Indonesian officials and representatives from international organisations (World Bank, WFP, UN-CAPSA), research centres (ICASEPS) and universities (Bogor Agricultural University).

The authors would like to thank Peter Warr (Australian National University), Arief Anshory Yusuf (Padjadjaran University), and Hiroaki Ogawa (Osaka University) for their substantial contributions to the Indonesian case study.

# Table of contents



## Tables

## Figures

**Boxes**

# Abbreviations

| | |
|---|---|
| AFC | Asian Financial Crisis |
| AMIS | Agricultural Market Information System |
| APEC | Asia-Pacific Economic Cooperation |
| APTERR | ASEAN Plus Three Emergency Rice Reserve |
| ASEAN | Association of Southeast Asian Nations |
| BLT | Bantuan Langsung Tunai (Unconditional Cash Transfers) |
| BPH | Brown planthopper |
| BPS | Badan Pusat Statistik (Statistics Indonesia) |
| BULOG | Badan Urusan Logistik (Bureau of Logistics, Indonesia) |
| CASERD | Center for Agro-Socioeconomic Research and Development |
| CRED | Centre for Research on the Epidemiology of Disasters |
| FAO | Food and Agriculture Organisation of the UN |
| GHI | Global Hunger Index |
| IDR | Indonesian Rupiah (1000 IDR= 7.8 US cents, 16.02.2015) |
| IFAD | International Fund for Agricultural Development (agency of the UN) |
| ICFC | International Finance Corporation |
| IFLS | Indonesian Family Life Survey |
| IFPRI | International Food Policy Research Institute |
| IMF | International Monetary Fund |
| kcal | kilocalories |
| MDG | Millennium Development Goal |
| OFDA | Office of US Foreign Disaster Assistance |
| PARM | Platform for Agricultural Risk Management (IFAD) |
| PPFS | Policy Partnership on Food Security (APEC) |
| SUR (SURE) | Seemingly Unrelated Regression (Estimator/Estimation) |
| SUSENAS | Survei Sosial Ekonomi Nasional (National Socioeconomic Survey) |
| UN | United Nations |
| UNCTAD | United Nations Conference on Trade and Development |
| UNHLTF | High-Level Task Force of the UN Secretary-General |
| WFP | World Food Programme |
| WHO | World Health Organisation |
| WTO | World Trade Organisation |

# Executive summary

Many recent concerns about food security focus on unpredictable but shorter-lived threats to current food security levels such as price shocks and natural disasters. Unlike chronic food insecurity, transitory food insecurity occurs because of a temporary decline in household access to adequate food. Shocks like droughts or economic downturns can affect individuals who normally have appropriate access to food, threatening the stability of food security which implies adequate access to food at all times. It is particularly relevant for emerging economies that are rapidly reducing poverty and the prevalence of food insecurity, but are still vulnerable to shocks that could bring transitory food insecurity. Developed countries also sometimes raise these concerns when justifying their agricultural policies.

This report presents a risk-management framework for addressing transitory food insecurity. It is designed to examine the robustness of policy responses to managing the risks and uncertainty associated with various threats to food security. This framework is then applied to assess the risk of food insecurity in a large emerging economy, Indonesia, and to identify robust policy responses to food insecurity risks.

The analytical framework comprises three steps: *preparatory analysis*, *risk assessment* and *policy analysis*. The *preparatory analysis* is technical in nature and involves identifying data sources, including household expenditure surveys, indicators and models. A consultation process is proposed for the second and third steps. *Risk assessment* draws on the perceived food security threats of experts and stakeholders and on available scientific and statistical evidence. This evidence is then transformed into a set of scenarios, each one corresponding to a specific perceived food risk, thereby allowing a rigorous assessment of its likelihood and its estimated impact on food security. The participation of experts and stakeholders is a key factor in identifying plausible food insecurity scenarios. The subsequent *policy analysis* focuses on existing and potentially new policy instruments, and their impacts on each scenario. A portfolio approach is then adopted to analyse policies and scenarios jointly.

Following a consultation process among stakeholders and policy makers, five scenarios were selected as major threats to food security in Indonesia: a price hike in the world rice market, a macroeconomic crisis, an increase in the world energy price, failure of the rice crop due to a pest infestation, and an earthquake on the island of Sumatra. This list is not exhaustive, but it allows a significant number of risk situations to be represented in the subsequent assessment of risk and identification of policy options. The risk assessment shows that domestic economic and natural disaster scenarios are more important than global price hikes, both in terms of their likelihood and their potential impact on food insecurity. This fact should guide policy design; in particular, it highlights the need for early warning systems and disaster management strategies in Indonesia.

Indonesia's new Food Law No. 18/2012 endorses self-reliance (*kemandirian pangan*) as the guiding principle of food security and establishes domestic production of staples as the top priority. The 2010-14 strategic plan established production targets for 39 products. For five food commodities (rice, corn, soybean, sugar and beef) the targeted levels represent self-sufficiency. Price support policy for rice is largely implemented through a combination of direct intervention in the domestic market, including delivery of rice at subsidised prices to poor households (the Raskin programme), and trade restrictions. Due to these policies, the domestic rice price was 60% higher than the reference international price in 2010-12 compared to 8% higher in 2000-02. Household expenditure data (SUSENAS) reveal that current rice policies increase the rate of undernourishment in Indonesia by between 2 and 22%, depending on the degree of price transmission from international markets.

The performance of existing agricultural and social policies – the rice price support measures, the Raskin programme, the social programme of unconditional cash transfers (BLT), and fertiliser subsidies – was examined in each of the five selected risk scenarios. Other potential policy options were also investigated: a stylised crop insurance programme, a food aid programme distributing vouchers for staples, and a more targeted version of BLT.

The analysis shows that some policies that are positive in one scenario may have negative food security impacts in other scenarios, stressing the need for consistency of the policy strategy as a whole. Moreover, the current rice price support measures in Indonesia do not contribute to improve any dimension of food security, including stability, but instead they worsen the situation. A policy strategy that concentrates on addressing a single source of risk, such as a price spike in international markets, may increase vulnerability to other sources of risk such as domestic crop failure. More concretely, export restrictions can help avoid a surge of undernourishment only in the case of a price spike, which is estimated to occur just once in 30 years. Import restrictions worsen the food security situation in all other scenarios, in particular in the crop failure scenario, increasing the prevalence of undernourishment in Indonesia by 12 percentage points. Furthermore, the performance of Indonesia's social assistance effort in managing food insecurity risks could be improved by targeting the Raskin programme using food vouchers or cash transfers.

The BLT programme is slightly better targeted than Raskin, but there is scope here too for improved targeting. If these transfers were aimed at the poorest 20% of the population, their impact on reducing the overall rate of undernourishment would double. Fertiliser subsidies are not effective in reducing food insecurity in any scenario due to low income transfer efficiency, poor targeting and a weak impact on food prices. Crop insurance is expensive and difficult to develop among very small producers, while its contribution to food security is only positive for crop failure scenarios. Alongside these specific policy measures, strategic investments in people (education, training, extension services) and in physical infrastructure aimed to enhance innovation for long-term growth in agriculture are fully complementary with food security objectives and deserve further study.

There are six specific policy recommendations that emerge from the analysis of the risks to Indonesia's food security. The analysis suggests that it would be beneficial for the food security situation to:

- *Dismantle the rice subsidy programme Raskin and replace it with a food voucher programme* that, for the same cost, could be better targeted to the most vulnerable segment of the population. Food vouchers would be used to buy food staples consisting not only of rice but including other basic items. The exact list of products covered by the voucher should be decided in consultation with regional groups and could be differentiated regionally to respond well to local food preferences.

- *Improve the targeting of unconditional cash transfers*, for example by including triggers based on income and possibly special provisions based on weather or production losses for farmers. The convergence of social and food aid programmes should continue, and the new food voucher programme should be jointly managed with other social programmes to improve effectiveness and enable better monitoring of results.

- *Reform BULOG, by reducing its commercial activities and re-focussing its activities on the neutral management of emergency food reserves*. The floor purchasing price of rice should be phased out over time. Further analysis should be undertaken to define a good governance structure for the emergency reserve system and the links with the sub-national reserves and ASEAN+3 emergency rice reserve system (APTERR).

- *Reform the administrative requirements for agro-food imports, including import permits for rice*. Facilitating imports and the active participation of the Indonesian and foreign traders and investors can contribute to rural growth, incomes, and food supplies.
- *Promote a coordination agreement within ASEAN to restrain the use of export restrictions* and eliminate the administrative requirement of export permits. Export restrictions are very damaging for global and regional food security and, when prices increase, they can create policy traps that can exacerbate price spikes. The ASEAN region includes large exporters and importers of rice. More open and reliable regional trade among these trading countries could conceivably do more to reduce the variability of rice prices and ensure availability in all countries.
- *Phase out fertiliser subsidies and use the released budgetary amounts for strategic public investments,* including investment in people (education, training, extension services) and in physical infrastructure. Priorities should be identified in consultation with regional groups.

# *Chapter 1*

## Food insecurity: Concepts, measurement and causes

*This chapter reviews the concept and definition of food insecurity, its various dimensions and how it has been measured in different contexts.*

## 1.1 Introduction

The challenge of eliminating hunger and malnutrition has received increased global attention since the food price spikes and volatility of 2008. Meeting that challenge and overcoming hunger and malnutrition in the longer term is more about raising incomes of the poor than about food price fluctuations or other specific sources of risk (OECD, 2013a). At the same time, unpredictable events such as those of 2008 can in the shorter term have dire consequences for the food security of large numbers of people who are not perceived as trapped in chronic undernourishment.

Many of the recent concerns about food security relate in fact to perceived threats to current levels of food security, due for instance price shocks and natural disasters. These threats focus on the stability dimension of food security, so-called transitory food insecurity. This publication explores effective policy responses to this risk environment as part of a broader policy strategy designed to improve availability, access and utilisation of food. A case study on Indonesia highlights the issues involved.

The report is organised as follows. This chapter reviews the concept and definition of food insecurity, its various dimensions and how it has been measured in different contexts. Chapter 2 presents an analytical framework intended as a tool for designing policies to deal with the risks of transitory food insecurity. The point of departure for this framework is the key idea that risk is a defining characteristic of transitory food insecurity; hence, an evaluation of the risks of different types of food insecurity must precede the analysis of the performance of different policy instruments in dealing with it. Guidelines for the application of this analytical framework in three stylised steps are also presented. The underlying formal risk model is set out in Annex 2A.

Chapter 3 covers the implementation of the first step (*preparatory analysis*) in the Indonesian context. This step involves identifying the available data bases and modelling tools needed to analyse the relevant policy options. The current food security situation in Indonesia is described, drawing on the available data sources. Two annexes follow this section. Annex 3A gives details of a LA-AIDS household demand system estimated for Indonesia for this study, while Annex 3B outlines an existing general equilibrium model INDONESIA-E3. Both models are used in the policy impact analysis.

Chapter 3 also reports on the second step (*risk assessment*) that was undertaken by means of a consultation process among experts and stakeholders, together with a review of available scientific and statistical evidence. The main threats to food security stability are identified and are then summarised in the form of five stylised scenarios. Annex 3.C presents some additional scenarios that were identified by experts but not retained right through to the final policy dialogue stage of the analysis. The results of the policy analysis are also presented in this chapter; it examines and compares the simulated impacts of Indonesia's main agricultural and food policy instruments, plus several hypothetical instruments not currently in use, in both the reference scenario (absence of a transitory shock to food security) and in the five scenarios involving a specific food security shock.

Chapter 4 draws on this analysis to evaluate the performance of the current portfolio of policy instruments for dealing with transitory food insecurity, and makes a number of policy recommendations.

## 1.2 A global concern

The recent instability of international food prices and supplies has increased global awareness about hunger, and governments and political leaders have expressed their concerns about global food security and the risks of food insecurity (Box 1.1). Food security is the core focus of international organisations such as the Food and Agriculture Organisation (FAO) and the World Food Programme. According to a recent United Nations report, the world as a whole is on target to achieve the first Millennium Development Goal target of halving the proportion of the world population suffering from hunger, but still immediate additional efforts are required, especially in countries which have made little headway (UN Millennium Development Goals Report, 2014).

**Box 1.1. Increasing global awareness of food security and the management of its associated risks**

At the OECD Ministers meeting in Paris in February 2010, priority was given to "an integrated approach to food security... involving a mix of domestic production, international trade, stocks, safety nets for the poor and other measures". This integrated approach seeks to avoid situations of food insecurity caused by a variety of circumstances.

The world commodity price spikes of 2008 and 2011 raised international interest in price volatility and its impact on food security. At the G20 summit in November 2010, leaders requested FAO, IFAD, IMF, OECD, UNCTAD, WFP, the World Bank and WTO (and later IFPRI and UN HLTF), "to develop options... on how to better mitigate and manage the risks associated with the price volatility of food... to protect the most vulnerable". The resulting report, *Price Volatility and Agricultural Markets: Policy Responses,* was co-ordinated by FAO and OECD and was a main input to G20 discussions in June 2011. A portfolio of possible instruments to manage diverse price and production risks, including different types of insurance, future contracts, options and co-operative solutions were outlined. Several ideas from this report were reflected in the June 2011 Ministerial declaration "Action Plan on Food Price Volatility and Agriculture".

The Action Plan recognises that "managing the risk... in developed and developing countries would provide an important contribution to... strengthen food security". An important outcome was the creation of the Agricultural Market Information System (AMIS), an initiative designed to enhance food market transparency and foster co-ordination of policy action when responding to international price volatility.

The Action Plan also argues for including risk management when developing policies, and for developing an "Agriculture and Food Security Risk Management Toolbox" for vulnerable countries, firms and farms. Annex 5 of the Action Plan expands on the idea of this toolbox, focusing on the need to integrate risk assessment into agricultural development programmes and on "government level risk" tools such as the New Price Risk Management (APRM) product of the IFC, counter-cyclical instruments for vulnerable countries, and hedging strategies for humanitarian agencies and the WFP. A risk management toolbox is proposed to address food security concerns including instruments for farmers, such as insurance, and for governments. It also highlights the importance of risk assessment for development and food security policies.

Two follow-up reports on risk management were issued for the September 2011 financial/development G20 ministerial: a progress report on the risk management advisory mechanism and a report on the state of play of multilateral and regional development banks (MDB, 2011). The latter report focused on "government-level risk management solutions" and followed "a holistic approach to risk management". It reported on a list of financial instruments provided by the MDB to help government manage risks, such as on currency exchange rate, commodity prices, and catastrophes.

The June 2012 Los Cabos declaration of G20 leaders, under the heading of enhancing food security, endorsed the initiative to create the Platform for Agricultural Risk Management (PARM). The G20 development group meeting in February 2013 endorsed the conceptual note of PARM drafted in collaboration between several international organisations (including OECD) and development agencies. The approach emphasises risk assessment and a holistic approach. PARM began to operate in 2014, hosted by the International Fund for Agriculture Development (IFAD) in Rome.

Eliminating hunger and malnutrition is one of the most intractable problems humanity faces. OECD (2013a) distilled from previous work the main priorities for ensuring long-term global food security and concluded that, while price levels matter, they are not the fundamental problem. The persistence of global hunger – the chief manifestation of food insecurity – is a chronic problem that pre-dates the current period of higher food prices. Indeed, there were as many hungry people in the world in the early 2000s, when international food prices were at an all-time low, as there are today.

Global food security is a complex chronic problem. Eliminating global hunger is more about raising the incomes of the poor than an issue of food prices (OECD, 2013a), and this challenge requires a combination of consistent, long-term policies oriented towards economic development and social protection. At the same time, the fears and concerns about increasing food insecurity as a result of price hikes or other shocks are also legitimate, since stability is another important dimension of food security, along with regular availability, access and utilisation. The policy challenge is how to respond to concerns about stability without compromising the long-term improvements in chronic food insecurity.

The experience of the 2008 international price spike has increased policy and public awareness about the potential impacts of these events on food security. In various emerging economies, food security is endorsed as a main objective of agricultural policies. This is particularly true as concerns rice markets and Asian countries. For instance, India, China and Indonesia have developed self-sufficiency policy objectives and some countries, including Indonesia, are expanding the scope of these objectives. Stressing the need to increase domestic production and promote self-sufficiency seems to imply that trade disruptions are perceived as the main threat to food security. Past experience has shown, however, that such an approach can result in agricultural policies that are dominated by distorting forms of support, including price support, which in turn threaten the food security of poor consumers. These negative consequences are particularly acute when food shortages occur for other reasons, like an economic downturn or a natural disaster.

Food security concerns are also reflected in action by regional organisations but with a different emphasis. The APEC Policy Partnership on Food Security (PPFS), formed in 2011, is a forum for discussing issues related to food security, bringing together individuals from the private and public sectors to help facilitate investment, liberalise trade and market access, and support sustainable development. The first component of the 2008 ASEAN Integrated Food Security (AIFS) Framework is labelled Emergency/Shortage Relief and focuses on stability, including food assistance programmes, diversification and the ASEAN Plus Three (ASEAN+3) Emergency Rice Reserve (APTERR).

This publication focuses on the stability dimension of food security and presents a framework for the rigorous assessment of a whole range of risks threatening food security in a given country. A portfolio of measures to deal with risks should include policies that are effective and efficient across different scenarios of food security threats. In essence, this means managing a whole set of threats in order to stabilise food security whilst at the same time not compromising efforts to improve chronic levels of food insecurity. Having said this, the present contribution does not address the core issue of eliminating structural poverty, and the need to improve economic development and access to health and sanitation. These latter issues are beyond the scope of the work reported here.

## 1.3 Definition of food security[1]

Amartya Sen, the 1990 Economics Nobel laureate, observed that "for many years, rational discussion of the food problems in the modern world was distracted by undue concentration on the comparative trends of population growth and the expansion of food output" (Drèze and Sen, 1990, p35). This Malthusian pessimism has not been vindicated by history, but it often comes back in public discussions on food security and, at the country level, can lead to policies aiming at self-sufficiency in food production. The broader entitlements approach proposed by Drèze and Sen does not specify a particular cause of famine, just a framework that identifies it as a widespread failure of entitlements for a substantial part of the population. The causes are diverse, including droughts, floods, general inflationary pressure, sharp recessionary loss of employment, war, and can occur even without a decline in food output or availability per head. Food production is not only a source of food supply, but for large sections of the population it is also the main source of livelihood. This explains why sudden falls in food output tend to go hand in hand with a collapse of entitlements.

These ideas were influential in framing the current FAO definition of food security, the main elements of which were formulated during the 1996 World Food Summit:

> *Food security exists when all people, at all times, have physical, social and economic access to sufficient, safe and nutritious food that meets their dietary needs and food preferences for an active and healthy life. Household food security is the application of this concept at the family level, with individuals within households as the focus of interest. Food insecurity exists when people do not have adequate physical, social or economic access to food as defined above. (*FAO, 2003).

This definition stresses both the *physical availability* of food and the *economic means* to procure it —the entitlement problem (Sen, 1981). Two further dimensions are also implicit in this definition: *utilisation* (appropriate diet and good food practices that maximise the nutritional contribution of food consumed), and *stability* to ensure that the first three conditions (access, entitlement, and utilisation) are met at all times and not merely on a periodic basis.

These four dimensions as defined by FAO (2006)[2] are set out in Table 1.1, which highlights that stability is a cross-cutting dimension of food security and hence requires managing the risk of interruption or deterioration in food availability, access or utilisation.

**Table 1.1. The four dimensions of food security in the FAO definition**

| Food | | Stability |
|---|---|---|
| Availability | The availability of sufficient quantities of food of appropriate quality, supplied through domestic production or imports (including food aid). | To be food secure, a population, household or individual must have access to adequate food at all times. They should not risk losing access to food as a consequence of sudden shocks (e.g. an economic or climatic crisis) or cyclical events (e.g. seasonal food insecurity). The concept of stability can therefore refer to both the availability and access dimensions of food security. |
| Access | Access by individuals to adequate resources (entitlements) for acquiring appropriate foods for a nutritious diet. Entitlements are defined as the set of all commodity bundles over which a person can establish command given the legal, political, economic and social arrangements of the community in which they live (including traditional rights such as access to common resources). | |
| Utilisation | Utilisation of food through adequate diet, clean water, sanitation and health care to reach a state of nutritional well-being where all physiological needs are met. This brings out the importance of non-food inputs in food security. | |

*Source*: Text from FAO (2006), Briefing Note "Food Security". Table adapted by OECD.

There are individuals or households that are chronically food insecure because of a continuously inadequate diet due to their structural inability to acquire food, usually due to poverty. There are other individuals and households that suffer a transitory or temporary decline in their access to adequate food. This temporary or transitory disruption can be due to shocks that affect income, assets or infrastructure and it has been denominated "transitory" food insecurity by the World Bank (1986). The use of this term highlights the different nature of the stability dimension of food security. The risk of transitory lack of access to food is due to potential shocks affecting individuals who normally have adequate access to food. This risk can be managed through appropriate strategies that allow the reduction or mitigation of their effects and their potential long lasting consequences. Long-term exposure to risk can cause households to lapse into chronic food insecurity (e.g. if they are forced to sell their assets).

The response to chronic levels of food insecurity should be medium- to long-term policies generating broad-based income growth with lasting impacts in reducing global hunger. Policies and investments that stimulate income growth are likely to reduce the need for short-term fixes that cope with consequences of low incomes but do not tackle the underlying causes (OECD, 2013a). Improving availability requires investing in productivity growth, innovation and infrastructure. Improving access to food requires social protection systems appropriate to the development stage of each country. Improving diets and utilisation requires education, information, direct nutrition intervention, and better access to health services and sanitation. The main driver of these three dimensions of food security is inclusive and sustainable economic development, supported by innovation, competitiveness and appropriate social policies.

By contrast, transitory disruptions that threaten food security have short- to medium-term emergency and stability dimensions. A country can find itself in an emergency situation due to a specific shock that requires a rapid response: a price spike of a staple food or sudden food inflation, a severe drought that produces crop failure, or an economic slowdown reducing the income of the poor, a local natural disaster like an earthquake that destroys assets and livelihoods. The framework set out in this report is not concerned with the management of disasters and emergencies themselves, but rather with the portfolio of policies that can respond to transitory shocks to food availability and access. Many countries are prone to potential events that could occur at any time and plunge large segments of the population into extreme food insecurity of unspecified duration. This defines the stability dimension of food security that is the focus of this report. Policies in place need to be capable of managing of potential risky scenarios that threaten food security due to various alternative causes, whilst also being compatible with a policy environment that stimulates income growth and that reduces food insecurity across all scenarios including those where there is no particular shock.

## 1.4 Indicators of food insecurity

Given the multidimensional nature of food security, it is not possible to monitor all four food insecurity dimensions simultaneously with a single indicator (Cafiero, 2013). Therefore, several indicators have been developed to measure different aspects of food insecurity (OECD, 2013a), most of which are based on estimates of the share of the population that is under a given threshold of a relevant variable, such as calorie consumption, food expenditure or anthropometric measures indicating child malnutrition. According to Sibrián (2009), the most appropriate variables to measure food insecurity include dietary energy consumption (i.e. food consumption measured in energy units, e.g. kcal) to measure food deprivation, expenditure or income available to meet dietary energy needs to estimate food poverty, and weight- or height-for-age to quantify child under-nutrition.

Poverty, undernourishment and malnutrition are links in a vicious circle. This is why the first of the Millennium Development Goals (MDGs) is to eradicate extremes of both poverty and hunger. The World Bank (2004) emphasises that poverty is a root cause of hunger. The website of the World Health Organisation (WHO) Global Database on Child Growth and Malnutrition points out that[3] "…malnutrition is frequently part of a vicious circle that includes poverty and disease. These three factors are interlinked in such a way that each contributes to the presence and permanence of the others. Socioeconomic and political changes that improve health and nutrition can break the cycle; as can specific nutrition and health interventions. Malnutrition usually refers to a number of diseases, each with a specific cause related to one or more nutrients, for example protein, iodine, vitamin A or iron. In the present context malnutrition is synonymous with protein-energy malnutrition, which signifies an imbalance between the supply of protein and energy and the body's demand for them to ensure optimal growth and function". The existence of this vicious circle is well known in the literature as is the statistical analysis attempting to quantify the links between poverty and malnutrition (e.g. Gebhart 1920; Radhakrishna et al., 2004). However, these studies also find significant differences between poverty and malnutrition in terms of their prevalence and trends.

The definitive choice of a particular indicator of food insecurity depends on the particular aspect of food insecurity one wishes to examine (Box 1.2). Inevitably, the ability of these indicators to capture transitory shocks to food security depends to a large extent on the availability of the relevant data and the frequency of their measurement.

The FAO undernourishment index, the indicator most commonly used to measure food deprivation, is a macro indicator of the prevalence of undernourishment defined as the share of the population that consumes fewer calories than the minimum daily energy requirement. The WHO publishes an annual index of underweight children under five for a large number of countries. This type of indicator is, however, not sensitive to short-term changes such as sudden increases in food deficiency for those most deprived, and is unable to capture transitory crisis or droughts (Cafiero and Gennari, 2011; Masset, 2011). In general, the indicators of food security that are able to capture transitory shocks must be based on specific national or local surveys with frequently gathered information on the distribution of a relevant variable across individuals. Household level expenditure surveys with a food consumption module and a panel of households that can be followed over time are the most appropriate information source to assess the extent of food insecurity across individuals and how it changes due to different transitory factors. Such surveys typically allow the distribution of expenditure and calorie intake to be analysed across the whole sample, and although they rarely include information on nutritional outcomes such as underweight, they can already provide a very detailed and representative picture of the food security profile of the population. This degree of detail in the distribution is not possible with the FAO method of estimating the number of undernourished persons based on an imposed log-normal distribution. The advantage of this latter method is that it can be applied to countries that do not conduct a household consumption survey. However, the two methods are likely to produce different absolute numbers of undernourished persons. To avoid a futile debate about the absolute numbers and focus the discussion on the variability, the undernourishment threshold can be calibrated and applied to the survey distribution to replicate the official rate of undernourishment estimated by FAO.[4]

The use of food insecurity indicators to measure "transitory" food insecurity is also challenging. At the individual or household level, the identification of chronic or transitory undernourished can only be done using panel data. Chronically food-insecure individuals or households are those that are below the threshold of a given indicator for several years in a row. Normally only a few years of panel data are available and this limits the time horizon for "chronic" food insecurity can be measured at household level. Transitory undernourished households or individuals in a given year are those that are below the threshold in that year, but above in other years in the sample.

**Box 1.2. Main indicators of food insecurity**

There are three types of indicator that are most commonly used to monitor food insecurity: child malnutrition, prevalence of undernourishment and food poverty. Prevalence of undernourishment, as measured by the FAO, is estimated as the proportion of people in a given population who do not consume enough food to meet their daily energy requirement to live a healthy and active life. The measurement method is set within a probability distribution framework. FAO assumes that food consumption expressed in kcal per capita per day is log-normally distributed. The parameters of the log normal energy consumption distributions are calibrated in each country from available data, most often food balance sheets for the average energy availability and household surveys for the variance. The percentage of undernourished population (also referred to as the chronically hungry) is measured as the area under food consumption distribution curve below the minimum requirement cut-off point measured in kcal/person/day, which is specific for each country. The FAO uses a related indicator to monitor the Millennium Development Goal of halving the number of undernourished. Another related Food Security Indicator in the FAO statistics is the depth of food deficit, which measures (in kcal) the gap between the average dietary energy consumption of the undernourished population (food-deprived) and the average dietary energy requirement, scaled up to the population as a whole by the total number of food-deprived persons. This indicator captures both the availability and access dimensions of food insecurity. Other similar indicators can be developed focused on a minimum intake of different nutrients (e.g. proteins) rather than energy (see NutVal.net, an initiative from the World Food Programme WFP and the UN Refugee Agency UNHCR).

Food poverty is defined as the proportion of the population living on less than the cost of their dietary energy needs. The cost of the dietary energy needs — the food poverty line — is estimated based on the assumption of the energy requirement for a healthy life of a representative individual in a given population (usually 2 100 calories per capita per day) and the cost of a food basket required to meet that energy requirement, taking into account the need for a properly balanced diet and the food habits of a given population. The share of the population whose expenditures (or income) fall under the food poverty line is considered food poor. This indicator captures the access dimension of food insecurity.

Child malnutrition is typically assessed through anthropometric indicators, which are related to body size and composition. Each indicator is compared to a reference population to establish whether a child is malnourished or not. There are three types of indicator: weight-for-age to measure underweight, height-for-age to measure stunting and weight-for-height to measure wasting. A child is considered malnourished, or more specifically under-nourished, if any of these indexes is lower than the threshold level established based on the values of these indicators in the reference population. The WHO's index of the prevalence of underweight is estimated as the proportion of children aged 0-5 years whose weight-for-age falls more than two standard deviations below the median of the reference population. Given their focus on nutritional status, the child malnutrition indicator measures the consequences of food insecurity regardless of the underlying reason for food insecurity: availability, access and (or) utilisation. The FAO publishes related indicators it its Food Security Indicators database under the "utilisation" heading.

IFPRI has developed a Global Hunger Index (GHI) that combines with equal weights *undernourishment, child underweight* and *child mortality*.

*Source:* Cafiero (2013), FAO (2008), Ravallion and Bidani (1994), Sibrián (2009), Setboonsarng (2005), IFPRI (2013) and OECD (2013a).

## 1.5 Stability of food security at household and government levels

The FAO definition of food security covers two levels of governance of food security: a household level and a government level. At the household level, members of the household must have access at all times to sufficient, safe and nutritious food. Various events, from unemployment to price spikes, can put at risk the availability, access and utilisation of food at this micro level. Among (farm) households, net sellers and net buyers of food face different risks. Risk management strategies can help both farm and non-farm households to manage their risks (OECD, 2013b).

Household risks are mainly related to potential shocks in purchasing power and income, including shocks in production for farm households. These shocks are managed at the level of individual households where income and assets can typically be pooled, and it is thus only at this level that an appropriate assessment of risks be made. Income diversification and market risk management tools are potential options that can be used by households. The work reported in OECD (2013b) analyses these aspects for farm households, but additional issues related to food consumers in rural or urban areas are raised for non-farm households (see Table 1.2. below). Following the OECD agricultural risk management framework (OECD, 2009; OECD, 2011), it is the responsibility of the household to manage its risks, including food insecurity risk using household, community (Townsend, 2013) or market instruments. However, it is beyond the capacity of households to cope with catastrophic events (rare, damaging and systemic); here, government has the responsibility of ensuring that they do not generate high food insecurity, and of facilitating households' access to efficient risk management tools.

At government level, policies are implemented to improve both the "chronic" level of food insecurity and the stability of food security in the face of typical risks. Most societies demand an acceptable level of food security that they can afford, and the stability objective requires managing different threats to this minimum acceptable level. The set of policies must be able to respond to risk from different sources, from high world market price volatility to domestic crop failure. In short, risk assessment and management at the household level is crucial, but a stable aggregate food security environment is also needed, and there are also important interactions between aggregate and individual vulnerability of rural and urban households that policy makers need to consider. In the current climate of heightened awareness of food insecurity risk, some countries may need to consider enhancing the linkage between risk assessment and management at household level and government policy at the macro level.

The stability dimension of food security at the government level implies managing risks of transitory increases in the total number of undernourished. This is the main focus of the framework for the analysis of transitory food insecurity presented in this study. Taking food availability as an example, risk comes from a shortfall in food supply due, for instance, to a significant crop failure. Government policies need to be able to manage, in a coordinated way, food supplies coming from production, imports and stocks so that the whole set of trade and agricultural policies play a role in responding to availability risks. As for access, it is the role of the government to enhance households' access to efficient risk management strategies, whether through economic diversification or the use of community or market insurance or, in the most extreme cases, public safety nets that can be scaled up in the case of a crisis. Finally, as for utilisation, direct nutritional intervention may be needed.

Risk management techniques and strategies are appropriate tools for analysing and managing situations of perceived risks or threats to stability of availability, access and utilisation of food, and they can also contribute to development and income growth objectives. The first step in analysing the management of risks threatening food security at government level is a rigorous assessment of food insecurity risks facing the country. Sources of risk need to be identified and analysed, and probabilities and correlations need to be assessed. A portfolio analysis will typically require a diversified set of policies that can respond to a variety of events or scenarios. However, consistent policy design requires the impact of each policy across *all* risk scenarios be examined, including its impact on the level of chronic food insecurity and poverty. A typical example of an inconsistency to be avoided is a tariff on a staple food that is supposed to increase self-sufficiency and reduce the perceived food security risk from trade, but which also permanently increases the cost of food for the poorest consumers.

Food insecurity is perceived as a major risk in many countries, and some OECD countries argue that food security objectives motivate their policies.[5] The perception of food insecurity is usually the result of a general fear of potentially devastating and catastrophic effects that such insecurity could have in the country concerned or on a global scale. Elsewhere, it is likely that food security will remain a major priority for many governments when designing or reforming their agricultural policies, particularly in emerging economies in Asia that have increasing resources to spend on their relatively shrinking agricultural sectors.

**Table 1.2. The stability dimension of food security at household and government**

Examples of main risks[1] and potential strategies

| | | Farm households: Producers | Non-farm households: Consumers | Aggregate or national |
|---|---|---|---|---|
| Stability of AVAILABILITY | *Threat:* | *Insufficient farm production* | *Lack of food in the markets* | *Insufficient production + imports + stocks* |
| | *Potential strategy:* | Improved techniques, diversification, risk management | Storage, food aid | Trade and agricultural policies |
| Stability of ACCESS | *Threat:* | *Low food prices or insufficient income* | *High food prices or low income (e.g. unemployment)* | *Lack of entitlements for a section of population* |
| | *Potential strategy:* | Income diversification, market risk management tools, cash transfer, food aid | Income diversification, market risk management tools, cash transfer, food aid | Enhanced risk management tools and safety nets |
| Stability of UTILISATION | *Threat:* | *Lack of food safety / unhealthy diets*<br>*Lack of access to potable water* | | |
| | *Potential strategy:* | Schooling, education, investment<br>Direct nutrition interventions | | |

1. The risks in this table do not refer to trends, structural issues or threats to these trends and structural features.

It is rare, however, that a government's pursuit of food security objectives is based on a rigorous assessment of the likelihood and circumstances of food insecurity threats. The risk of food insecurity typically remains undefined or loosely defined. There are nevertheless some countries, such as the United Kingdom, that have engaged in a systematic assessment of threats to its food security.[6] Risk assessment at this aggregate level needs to involve not only experts, but also government and stakeholders. Their perceptions are the main drivers of policy choices and need to be surveyed and contrasted with the available evidence, as well as the views of the experts. This participatory approach facilitates the evaluation of such risks, underpins the communication of their likelihood and potential damage, and enhances the understanding of the set of measures and strategies to manage the risks of food security. Effective risk assessment cannot be done from an external, distant perspective, but requires the collaboration of those directly affected by food security threats. A crucial step in the framework proposed here consists in identifying scenarios or events that represent a threat to food security or a food insecurity risk, and to assess the likelihood and impact of each scenario. Table 1.3 provides examples of the wide range of threats, including macroeconomic risks that are not discussed here but should be considered in a full risk assessment process. Some of those risks include conflict and war, which are hard to predict, as well as being very difficult to prevent or cope with, and hence particularly difficult to include in this analytical framework.

Each threat can be represented by a stylised scenario that describes a real food insecurity risk that can be documented with rigorous scientific information concerning its likelihood and eventual impact. Each scenario needs to be understood as a category of events rather than a single event, covering a range of similar threats. It is a simplified representation of perceived risks with the purpose of facilitating assessment and a rigorous dialogue about best policy responses.

Once risks have been fully assessed, the policy analysis can identify possible policy responses and their impact in different risk scenarios. The analysis of household data is very useful for quantifying the impact of different scenarios and policy responses on household food insecurity.

**Table 1.3. Examples of threats to food security**

| | Political | Technical | Demographic/ economic | Environmental |
|---|---|---|---|---|
| Availability | Wars, export restrictions, embargoes, breakdown of international trade | Inappropriate farming practices | Population growth, increased demand, high world prices, difficulties in the balance of payments | Floods, droughts, plant and animal diseases (increased by climate change?) |
| Access | Civil conflicts, government restrictions | Lack of transport | Economic downturn, unemployment, food inflation | Extreme weather events |
| Utilisation | Regulatory failures | Contamination | Longer supply chain | Pest and diseases |

*Source*: Adapted from DEFRA (2010).

# *Chapter 2*

## Analytical basis for the transitory food insecurity framework

*This chapter presents the theoretical basis for analysing the stability dimension of food security at government level. First, the target variables defining food security are identified using the distribution of individuals or households. The food-insecure segment of the population is defined as those below a given threshold, and the instability of food security is analysed based on the distribution of prevalence of food security across different scenarios and risks. Second, the type of information required to define a scenario is discussed using stylised examples. The risk assessment process should lead to estimates of the likelihood of the threat and its impact on the prevalence of food security. Finally, a portfolio approach to policy decisions is explained with stylised examples. As part of the process, the effectiveness and efficiency of different policy measures in different scenarios need to be analysed and reconciled with the government's attitude towards avoiding the risk of lower food security in different scenarios.*

## 2.1 Measuring instability of food security

The standard indicators of hunger and food security (see section 1.4. and Box 1.2.) are based on the prevalence of undernourishment, of under-nutrition in children or of food poverty (Ravallion and Bidani, 1994; Sibrián, 2009; OECD, 2012a). The definitions of food insecurity are in terms of the lower tail of the distribution over individuals of the indicator in question, be it nourishment (dietary energy consumption), children's weight, or per capita food expenditure in a given country. The tail of the distribution is defined in relation to a threshold value of the chosen indicator, which defines the minimum value of the indicator that is compatible with the definition of food security. The choice of which indicator to use for measuring food insecurity depends on the particular dimension of food insecurity of interest.

The graphical presentation in Figure 2.1 relies on a distribution of the relevant food security indicator, which may be based either on empirical information obtained from a household survey for the population concerned, or (where this is not available) an assumed distribution such as the lognormal. Figure 2.1 illustrates the approach for the indicator *per capita dietary energy consumption*, where the distribution is scaled such that the area under the entire curve sums to 1.[7] The threshold value μ denotes the minimum per capita energy consumption consistent with food security, for example 1 800 kilocalories per day (which will be specific to the circumstances of a given population). The *prevalence of food insecurity* "I" in a given country (or region) is defined as the share of the population whose calorie intake is below this threshold; that is, the area under the tail of the distribution in Figure 2.1, often known as the rate of undernourishment. The *prevalence of food security* "S" is defined as the share of the population above this threshold, with S=1-I.

Current levels of food insecurity can be due to chronic, structural causes or to more transitory ones. In the first case, the position of the distribution relative to the threshold value will be relatively stable over time. The greater the proportion of food insecurity that originates from transitory shocks, the more likely it is that there are horizontal displacements of the distribution from one measurement period to the next. For example, after a negative income shock, the whole distribution of energy intake will move to the left, leaving a larger segment of the population under the food insecurity threshold. The prevalence of food security "S" in that country will be reduced and food insecurity "I" increased. The distribution in Figure 2.1 may also take different shapes after different shocks or scenarios that modify the pattern of food consumption in the population. Each scenario and shape will lead to a different rate of food insecurity or undernourishment.

In theory, expenditure and consumption surveys of a household panel can be used to decompose undernourishment into chronic and transitory components. However, since these surveys are normally undertaken once a year and refer to a specific time during the year, they cannot capture seasonal variations in food insecurity or episodes of food insecurity of duration less than one year. For these reasons, the term "transitory" cannot in practice be defined unambiguously with reference to duration, but depends on the frequency of observation of the indicator variable or variables.

Whereas Figure 2.1 shows the distribution of an indicator variable over the population at one point in time, and allows the measurement of the prevalence of food security, S, at that point in time, Figure 2.2 by contrast depicts the distribution of S (which can take values between 0% and 100%) across *different* probabilistic scenarios or states of nature. Each point in the distribution reflects therefore the probability of a given level of food security occurring in the country.[8] The mode or most likely observable value of S could be very high (e.g. the figure assumes that the most likely level of food security in the population, regardless of the scenario, is 99% of the population being food secure) but it could also be low as in some developing or emerging economies (e.g. 87% in Indonesia).

**Figure 2.1. Prevalence of food insecurity and prevalence of food security**

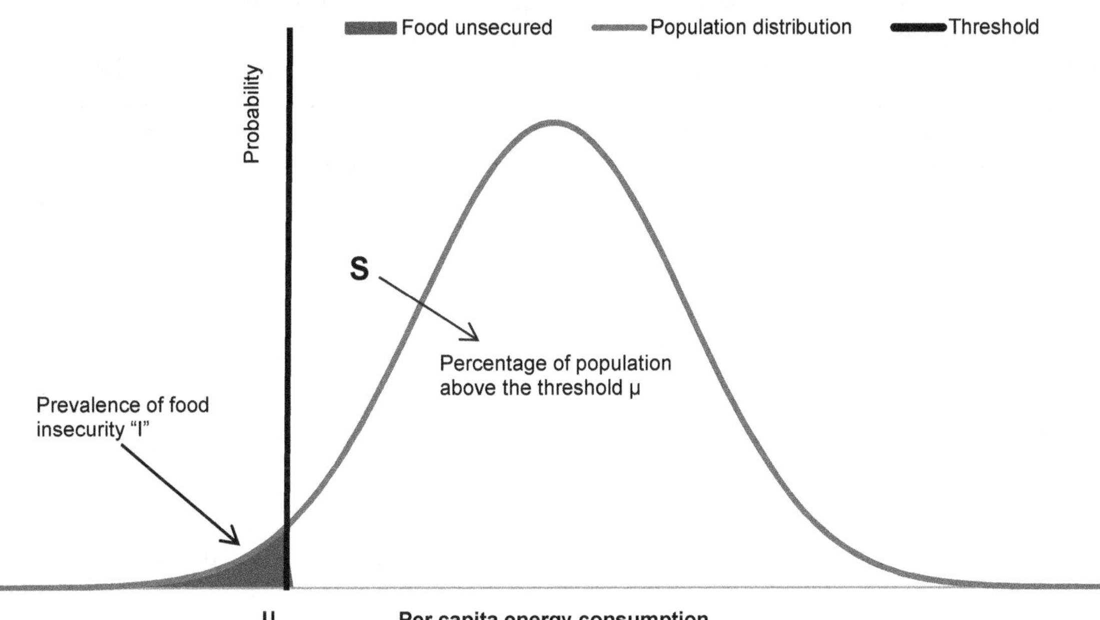


**Figure 2.2. Instability of food security**

Distribution of the food security indicator "S" across potential scenarios

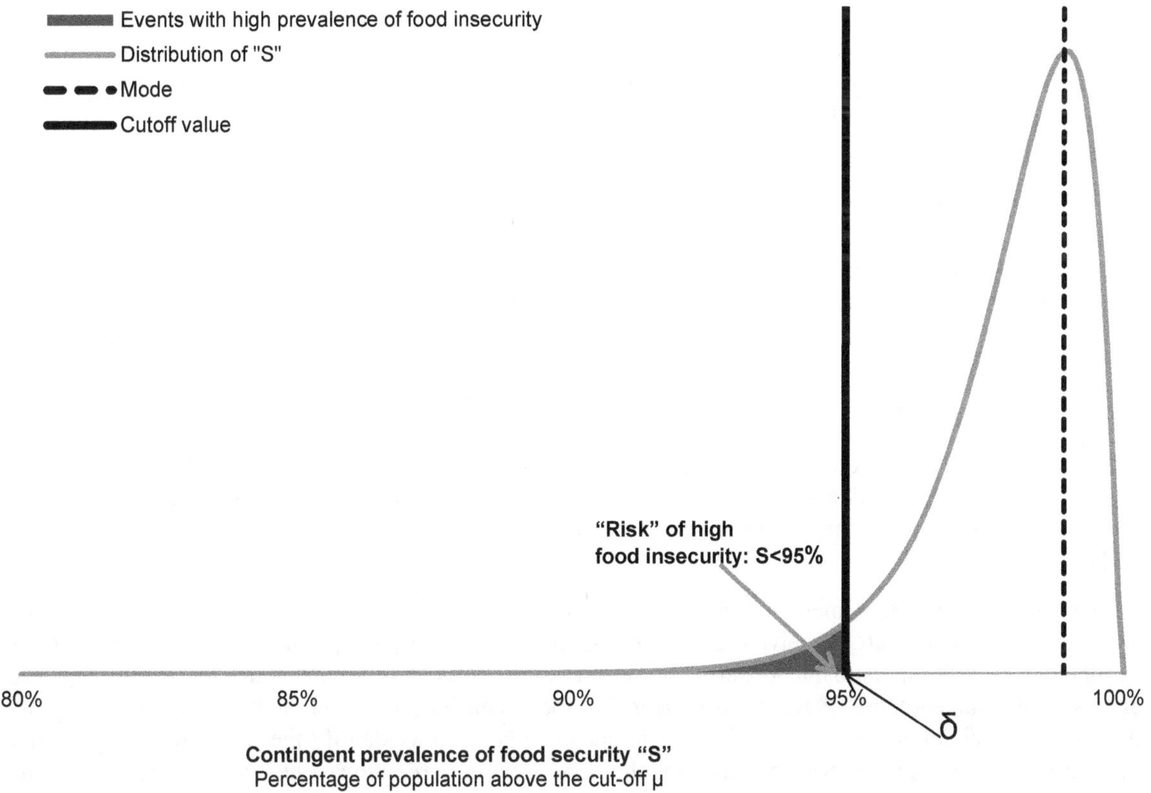

This graphical approach helps to visualise the two interlinked objectives of government food security policies: first, to shift the peak of the distribution of S to the right, so as to improve the expected rate of food security prevalence regardless of the conditions of a given year; and second, to reduce the dispersion of S, in particular to reduce the skew in the left-hand tail of the distribution, so that in extreme but less likely transitory situations with the potential to damage the food security of large numbers of households, the consequences for their food security are mitigated. The first objective matches that of reducing underlying chronic food insecurity in the population, whilst the second objective requires policies that can 'absorb' or deflect the negative consequences for food security caused by transitory shocks.

In Figure 2.2, the area underneath the distribution curve in the lower tail to the left of an arbitrary cut-off point of 95% (denoted by δ) is 3.5% of the total area under the curve; thus, the graph assumes that the prevalence of food security will fall below 95% with a probability of 3.5% (or once every 29 years) and implies a 5% prevalence of undernourishment in of the population. The value of δ can be set by the political process so as to represent the minimum socially acceptable level of food security prevalence. It follows that, in this case, government would be particularly concerned with managing the extreme outcomes that push S below the value of δ.

Shocks of various kinds can put at risk the nutritional situation of households that are above but close to the food insecurity threshold. Reducing the dispersion of the distribution of S implies designing policies that make these households more resilient to food insecurity risks. To do this, policy makers need to know which are the events and circumstances that prevail in the states of nature corresponding to the outcomes below the value of δ, and what is the profile of the households that fall below the food security norms in these states of nature.

Thus, a crucial first step in designing appropriate risk management policies is a rigorous risk assessment and a description of the risk characteristics of the events that threaten food security, in particular those that occur in the distribution's tail and which increase the prevalence of food insecurity in the country. In practice, the identification of these extreme events and the assessment of their probabilities can only cover a limited number of contingencies. Consequently, instead of a continuous probability distribution like the one depicted in Figure 2.2, the risk assessment process will lead to a table of scenarios with the corresponding probabilities and expected outcomes in terms of undernourishment.

## 2.2 Risk assessment of events threatening food security

The causes of food insecurity are typically surrounded by much uncertainty and many conflicting perceptions about the nature and probability of threats. Therefore, their risk assessment is no easy task, and needs to exploit all available information sources. The challenge of this risk analysis is to combine expert and statistical information with appropriate technical methods in order to obtain reliable estimates of two key parameters. First, the frequency or probability $\rho_i$ of occurrence of each scenario $i$. If a quantitative estimate of this probability is not possible, some qualitative assessment of likelihood would be required (e.g. likely, occasional, possible, unlikely, rare and remote). Second, the most likely impact in terms of prevalence of food security $S_i$. This latter assessment does not need to be fully quantitative, but for illustrative purposes, we assume that a full quantitative assessment can be undertaken.

A rigorous risk assessment should combine two sources of information of a very different nature about those events that are perceived as main risks, namely the risk perceptions of stakeholders and any available objective statistical and scientific information. Risk assessment starts with eliciting and analysing the stakeholders' perceptions concerning food security risks, such as the likelihood of future disasters and economic shocks. Different techniques can be used to identify the risks, including the use of questionnaires and interviews with stakeholders. These perceptions need to be contrasted and complemented with the available statistical data, including information from research data bases and models, and experts' assessment of the potential impacts of these scenarios on current levels of food security prevalence of food security.

The risks identified in this way then have to be formalised as specific events or scenarios, which are to be defined quantitatively if possible. Ideally, a discussion process with stakeholders, experts and policy makers will lead to a consensus on the set of scenarios that best define the perceived threats to food security in that country. The design of this process has to be adapted to the reality of each country with a view to ensuring the engagement of relevant stakeholders and the acceptance of the results of the process by participants, in particular, policy makers. Once the scenarios have been identified, they need to be validated by experts.

An example of five stylised scenarios for a hypothetical country is presented in Table 2.1. Each of Scenarios 1 to 5 is estimated to lead to high prevalence of undernourishment $I_i$, that is, low levels of food security with $S_i<\delta$. The scenarios that are assumed to define the food insecurity profile in this country are a food price spike, a food safety crisis, a catastrophic drought, and a trade disruption. The first row defines a reference scenario, that is, a 'normal' state of the world that has a high probability and will act as a benchmark for comparison with the other scenarios. A good risk assessment requires a description of these scenarios with as much precision as possible, ensuring they are informative and unambiguous profiles of plausible threats, and are identified and understood as such by stakeholders and the population in general. The second column of Table 2.1 includes a stylised description of the scenarios.

**Table 2.1. An example of a food insecurity profile: Set of events or scenarios that threaten current food security prevalence in a hypothetical country**

| Scenarios for identified risks | Description | Likelihood | Prevalence of undernourishment |
|---|---|---|---|
| Reference Scenario $R_1$ | "Normal" times with no particular negative shock affecting food security | $p_1$ | $I_1=1-S_1$ |
| Scenario $R_2$: Food price spike | High world and domestic prices impede poor domestic consumers' access to adequate nutritious food. | $p_2$ | $I_2=1-S_2$ |
| Scenario $R_3$: Food safety crisis | A food safety crisis is declared due to high incidence of bad quality food or a risk to human health | $p_3$ | $I_3=1-S_3$ |
| Scenario $R_4$: Catastrophic drought | Rainfall quantities remain well below average levels for more than one year, with record minimum food production. Food becomes scarce and, if the country is large or drought affects many countries, imports become expensive. | $p_4$ | $I_4=1-S_4$ |
| Scenario $R_5$: Trade disruption | Countries X and Y, traditional source of imports of the main staple decide to restrict their exports due to domestic circumstances in these countries | $P_5$ | $I_5=1-S_5$ |

## 2.3 Policy assessment

Experts and policy makers must then identify policy responses for the set of risks and scenarios. Different policy instruments can be more appropriate for different scenarios. An analysis of the impacts of the different policy options on the prevalence of undernourishment is required. If there is an expenditure and consumption survey in the country, the impact analysis will be extensively based on these data and on the use of associated economic models. The incidence of policy on undernourishment also needs an expert analysis: while cash transfers affect food consumption through income, price policies have a more direct impact on food demand. Cost-effectiveness analysis is also recommended. Table 2.2 presents an example of plausible policy responses to the food security threats and their consequences for undernourishment in each of the scenarios defined in Table 2.1. The first row presents the reference scenario for comparison, and all policies also have an impact in the reference scenario. The identification, description and quantification of the other policies and their impacts across scenarios are complex and require technical expertise combined with some policy modelling. The results of this technical work need to be presented in a simple way so that they are easy to understand and can be used in a policy dialogue with stakeholders.

The third to seventh columns in Table 2.2 contain estimates of the prevalence of undernourishment when the plausible stylised policies are applied to each scenario. Each policy is likely to improve this prevalence under at least some of the risk scenarios. However, each policy may not necessarily be effective in improving food security under other scenarios and, in particular, may or may not improve food security under the reference scenario. For instance, an investment in good animal, plant and human health services can be extremely useful for preventing the threat of a food crisis, but is not useful if the threat that materialises is a drought, and may make food more expensive in all scenarios, including the reference scenario. Tariff measures supporting self-sufficiency may aim at improving food security under the trade disruption scenario, but can have very negative impacts on food access by the poor in all other scenarios, and in particular in the reference scenario.

Quantifying the impact of each policy measure on undernourishment in each scenario is critical for a rational policy choice. However, no policy assessment will be able to quantify these effects fully and some estimates will remain qualitative. In general, a good policy response will require applying a portfolio of policy measures, including prevention, mitigation and coping measures, which could help to improve food security under a variety of food insecurity scenarios.

**Table 2.2. Matrix of policy responses to food insecurity scenarios and their impact on undernourishment: A hypothetical example**

| Scenario | Likelihood | Policy $P_1$: Main existing growth, development and social policies | Policy $P_2$: Investment in policy instruments for emergency cash transfers and targeted social policies | Policy $P_3$: Investment in animal, plant and human health services that prevent the pest and help managing the disease | Policy $P_4$: Engaging in trade agreements ensuring good access to a diversified source of imports; emergency stocks | Policy $P_5$: Supporting domestic production and self-sufficiency |
|---|---|---|---|---|---|---|
| Reference Scenario $R_1$ | $p_1$ | $I_1(P_1)$ | $I_1(P_2)$ | $I_1(P_3)$ | $I_1(P_4)$ | $I_1(P_5)$ |
| Scenario $R_2$: Food price spike | $p_2$ | $I_2(P_1)$ | $I_2(P_2)$ | $I_2(P_3)$ | $I_2(P_4)$ | $I_2(P_5)$ |
| Scenario $R_3$: Food safety crisis | $p_3$ | $I_3(P_1)$ | $I_3(P_2)$ | $I_3(P_3)$ | $I_3(P_4)$ | $I_3(P_5)$ |
| Scenario $R_4$: Catastrophic drought | $p_4$ | $I_4(P_1)$ | $I_4(P_2)$ | $I_4(P_3)$ | $I_4(P_4)$ | $I_4(P_5)$ |
| Scenario $R_5$: Trade disruption | $P_5$ | $I_5(P_1)$ | $I_5(P_2)$ | $I_5(P_3)$ | $I_5(P_4)$ | $I_5(P_5)$ |

A simple portfolio model can be used to analyse policy decisions in the context of a diversity of food insecurity scenarios. Table 2.3 provides a summary of the contingent outcomes associated with each policy as in Table 2.2. Results are presented in terms of the level of food security S that it is obtained from the prevalence of undernourishment (I) as S=1-I. Two summary indicators are given for each policy $i$=1,...5: the average level of food security across all the scenarios $\mu[S(P_i)]$, and its variability across scenarios measured by the standard deviation $\sigma[S(P_i)]$. The mean represents the expected level of food security under that policy, while the standard deviation measures the variability and serves as an inverse indicator of robustness. A robust choice is defined as one that performs "reasonably well" under a variety of different plausible scenarios even though it does not necessarily provide the highest expected level of food security (Antón et al., 2013). A robust policy choice avoids the worst outcomes. Another possible indicator of robustness could be the minimum value of food security across scenarios. Robust policies will have low standard deviations and high minimum values of S.

**Table 2.3. Impact of policy responses on the prevalence of food security across scenarios: An example**

| Scenario | Policy $P_1$: Main existing Growth, Development and Social policies | Policy $P_2$: Investment in policy instruments for emergency cash transfers and targeted social policies | Policy $P_3$: Investment in animal, plant and human health services that prevent the pest and help managing the disease | Policy $P_4$: Engaging in trade agreements ensuring good access to a diversified source of imports. Emergency stocks. | Policy $P_5$: Supporting domestic production and self-sufficiency |
|---|---|---|---|---|---|
| Average | $\mu[S(P_1)]$ | $\mu[S(P_2)]$ | $\mu[S(P_3)]$ | $\mu[S(P_4)]$ | $\mu[S(P_5)]$ |
| Standard Deviation | $\sigma[S(P_1)]$ | $\sigma[S(P_2)]$ | $\sigma[S(P_3)]$ | $\sigma[S(P_4)]$ | $\sigma[S(P_5)]$ |

Portfolio choice theory (Markowitz, 1952) can be applied to the numbers in Table 2.3. Each policy choice can be interpreted as a different portfolio of outcomes across scenarios. Some policies may dominate other policies both in terms of expected impact on food security (higher average) and robustness (lower standard deviation). Only the policies that are not dominated by other policies are efficient in the double sense of expected food security and robustness. The analysis of efficient policies in this double sense can already provide useful policy insights.

The application of the portfolio approach could be extended to allow the adoption of a policy set that combines several policy measures with different degrees of policy effort. A formal model of this is developed in Annex 2A. The model is characterised by a diversity of food insecurity scenarios with different probabilities and prevalence levels of food insecurity, a set of policy options with different budgetary implications and different capacity to improve food security under alternative scenarios, and a government that makes food security policy not only for the context of the most likely, reference scenario, but also takes account of the possibility of less frequent but more catastrophic scenarios that put food security particularly at risk. While the notion of policy optimality in this context is certainly too strong, the optimality condition in equation [1] in the annex provides several clear guidelines for good policy practices.

The theoretical analysis in this section provides a number of insights and observations, which are now discussed.

*Reference scenario as benchmark.* The reference scenario typically has the highest probability. The default policies for improving food security in this scenario target economic development and go beyond agricultural and social policies. They are likely to require the largest policy effort and budgetary expenditure. Marginal reallocations of the budget are likely to have relatively small impacts on food security in the reference scenario. This is why the reference scenario plays a pivotal role in defining the portfolio of policies. The policies that enhance food security in the reference scenario are likely to have positive effects on food security prevalence across several scenarios, plus positive effects on other policy objectives. The marginal gains from these policies could be taken as given and used as a benchmark for other policies that are more specifically oriented to food insecurity extreme scenarios. The cost of policies that reduce the impact on transitory undernourishment but increase chronic undernourishment are taken into account through this reference scenario.

*Both the probability of the scenario and the impact on food security matter.* More likely scenarios with higher probability $\rho_i$ deserve greater priority when designing food security policy. Typically, the reference scenario will be the most likely and hence the policies that promote food security in this scenario will receive priority. More unlikely or rare scenarios may appear to justify a smaller policy effort, but this is certainly not the case if in these more unlikely circumstances, food security is put at risk of being dramatically reduced with large numbers of undernourished. This is why the policy analysis needs to focus on both the most likely reference situation *and* the extreme shocks on food security.

**Figure 2.3. An example with two policies and two scenarios**

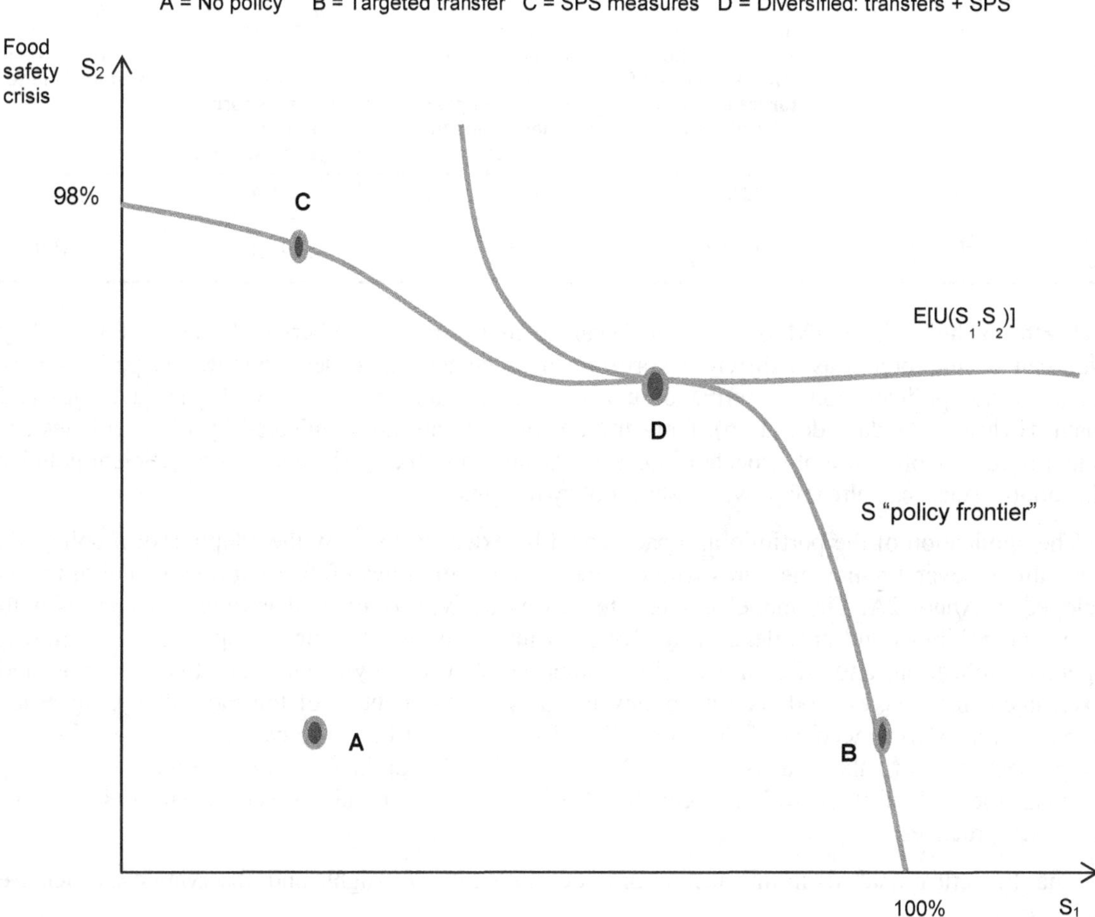


*Policies with similar impacts are redundant.* If two policies have similar positive impacts across scenarios, but differ strongly in their cost-effectiveness (with respect to budget costs), only the more cost-effective policy instrument will be used. Those policies that are effective across all possible scenarios are robust choices that deserve special attention. However, it is unlikely that two different polices will have identical impacts across all different scenarios.

*Policies with asymmetric impacts are complementary for diversification.* In the absence of a single policy that is robust across scenarios, policy instruments for each individual scenario are needed. If two policies have very different impacts across scenarios, then it will be rational to undertake a diversified policy strategy, using both policy measures to tackle food insecurity under different scenarios. For instance, if there are two potential policies that are effective only in one different scenario each, but fully ineffective in other scenarios, it may be optimal to use both in a diversified strategy.

This general framework can be illustrated with an example that combines two stylised scenarios and policies from Tables 2.2: the reference scenario and a food safety crisis, and targeted income transfers and a set of animal and plant health protection measures (sanitary and phytosanitary measures). Figure 2.3 depicts the policy frontier obtained as explained Annex 2A for these two food security scenarios and policies. Of course, targeted transfers to the poor can be effective and cost-effective in the case of the reference scenario, but they are likely to be ineffective with respect to a food safety crisis. In the same way, animal, plant and human health investment measures can be very effective against a food safety crisis, but fully ineffective in the reference scenario with no food safety crisis. In this case, a diversified strategy makes sense.

A very different illustrative example would be the result of combining the scenarios of trade disruption and drought in Table 2.2 ($R_4$ and $R_5$) and two policy responses: trade agreements to diversify imports, and self-sufficiency policies ($P_4$ and $P_5$). Supporting domestic production and self-sufficiency can be an effective policy to protect the country from the event of a trade disruption. But it will be ineffective and can aggravate the problem if the shock is internal and affects domestic production, such as a catastrophic drought. Free access to trade and diversified imports is a good policy strategy to protect against different kinds of domestic food insecurity shocks such as droughts that affect domestic production. However, unlike self-sufficiency policies, they can also contribute to reducing the risk of a trade disruption because the source of imports is diversified. This policy can be effective under different scenarios and its relative effectiveness needs to be analysed.

Policies are unlikely to be combinable in a continuum like those described Annex 2A and Figure 2.1. However, possible combinations of policies and their impact on food security across scenarios can be evaluated and their relative performance and dominance with respect to single policy options analysed.

## 2.4 Guidance for applying the analytical framework

These guidelines set out a process for applying the analytical framework to transitory food insecurity in Indonesia. The process follows three consecutive steps (Figure 2.4). First, preparatory work is needed to identify and analyse the available information sources for measuring the indicators of food insecurity and for quantifying behavioural responses to changes in the state of the world and to policy instruments. The second step is risk assessment, following the three-part conventional risk management approach comprising risk identification, risk analysis and risk evaluation. The final step consists of policy assessment and dialogue aiming to provide advice to the government.

**Figure 2.4. Application of the framework: A three-step process**

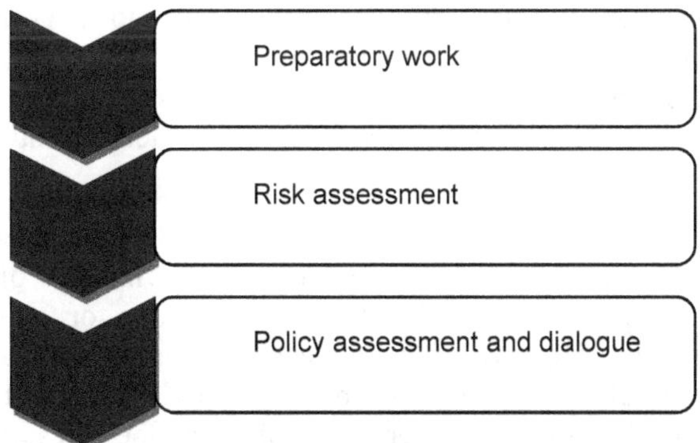

*Preparatory work*

The application of this framework in a given country has to be adapted to the reality of the data and information available. Thus, the first step involves inventorising the data sources are available, investigating how food insecurity can be measured using the available data, and developing tools to estimate the impacts of external shocks and policies (Figure 2.5).

**Figure 2.5. Step 1: Preparatory work**

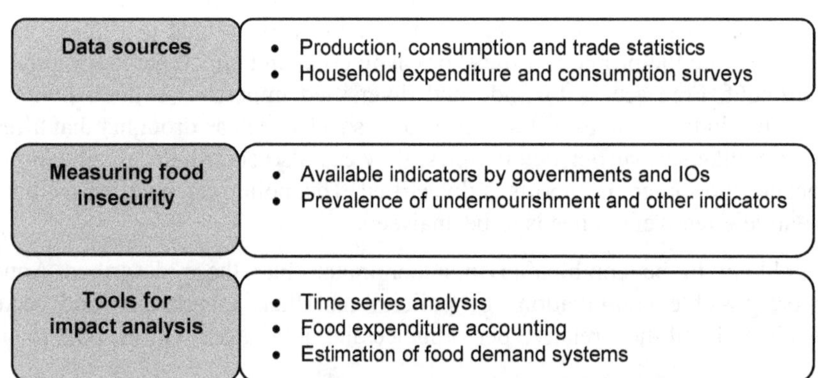


Aggregate data on food availability is a key information requirement. The main source of these data, when available, is the national household expenditure and consumption survey that many countries undertake.[9] This, and other sources of data, need to be investigated to make sure that at least one of the indicators in Box 1.2 can be estimated. Household expenditure survey data (like Indonesia's SUSENAS) can usually provide information on access to food and food expenditure, but does not normally have anthropometric information for indicators like child malnutrition.

The same survey data, complemented with other sources of information and analysis, are also useful for estimating behavioural reactions to different scenarios and policy impacts, particularly if there is a panel of households that can be followed for more than one year (SUSENAS has three-year panels, in which the same households participate for three consecutive years). These data could be exploited to obtain various relevant parameters and develop different types of tools. As discussed further below, a statistical analysis of past events could establish their correlation with food security outcomes, a simple spreadsheet exercise based on the consumption accounts would show the crude impacts of price and income changes, while the estimation of a full food demand system would permit the simulation of scenarios and policy changes.

A first potential use of the household survey is to estimate the impact of past events on the prevalence of food insecurity, and infer from these estimates the impacts of future events. The time series dimension of the data allows impacts to be estimated over time for the same households. This is, in theory, a powerful instrument. However, it is subject to significant difficulties and limitations in practice: the time series are rarely long enough to identify causal impacts, it is difficult to isolate the impact of specific external shocks from other events, and although the dates of the survey may be in the same year of a shock, there may be several months of delay with respect to the occurrence of the specific event. The extent to which survey data are useful for this purpose will depend on the specific event and the structure of the data set.

A second use of the household survey is to assess the impact of a given scenario defined in terms of variables, such as income, food prices and farm production that affect the structure of household expenditure. First-incidence estimates show how prices, income or production have a direct impact on the food consumption capacity of each household in the sample if the household continues its previous consumption pattern. Of course, this kind of exercise will typically overestimate the impact of scenarios and policy change on food insecurity because it assumes households do not adjustment their consumption patterns to the new situation.

Consequently, a more sophisticated method is preferable; whereby the database is used to estimate a complete demand system for each household. This could be done using the Almost Ideal Demand System (AIDS) or a similar approach (Deaton, 1997). The estimated demand system can then be used to simulate demand responses when prices, income, production or other variables affect the household consumption decisions. In the Indonesian case study reported below, this approach was used to simulate changes in the distribution of calorie intake in different scenarios and policy settings.

*A two-stage consultation process*

The preparatory work in Step 1 is followed by a consultation process with experts, stakeholders and policy makers. This consultation process is undertaken in two stages (Steps 2 and 3). Step 2 involves risk assessment and Step 3 consists of policy assessment and dialogue. In each of these two steps, there will be at least one consultation meeting with experts, stakeholders and policy makers.

Risk assessment in Step 2 (Figure 2.6) first requires the risks and shocks that can reduce food security to be identified. Both questionnaires and interviews with experts, stakeholders and policymakers could be used. In the Indonesian study, electronic questionnaires proved a useful tool for obtaining an initial overview of risk perceptions. Secondly, this information is used for a rigorous risk analysis that combines all additional sources of knowledge and expertise available to define a limited number of scenarios that respond to the risk perceptions, and to quantify their likely impact on availability and access to food. The data sources and models assembled in Step 1, and the input from local experts, can be very useful for this part of the exercise. Finally, a consultation meeting will help to validate and re-orientate those scenarios that will be selected for the scenarios report. The final set of scenarios that are retained needs to meet a delicately balanced set of criteria: the number of scenarios has to be small to facilitate the discussion with non-technical experts whilst nevertheless reflecting the diversity of the original sources of shock (economic, natural, domestic, international) and the diversity in the scale of the impact (regional, nation-wide); at the same time, the scenarios need to match, or at least be compatible with, the perceptions from the relevant stakeholders in the country whilst remaining consistent with scientific knowledge and statistical facts.

In Step 3 (Figure 2.6) the scenarios that were built during the risk assessment are confronted with different policy options. This takes the form of a rigorous policy analysis of the impact of a variety of potential policy responses on the food security indicator under each scenario. The role of experts and models is very important for giving credibility to the policy analysis that is ultimately submitted for discussion in a policy seminar with stakeholders and policy makers. The objective of this stage is the identification of policies or policy mixes that are robust across a diversity of scenarios. A final policy report is then submitted to the government for discussion.

**Figure 2.6. Two-step process or risk assessment and policy dialogue**

**Step 2. Risk assessment**

| | |
|---|---|
| **Risk identification** | - by experts and stakeholders<br>- questionnaires and interviews |
| **Risk analysis** | - by experts and analysts<br>- scientific analysis, statistics and modelling<br>- scenarios and impact on food |
| **Risk evaluation** | - risk assessment consultation<br>- SCENARIOS REPORT |

**Step 3. Policy dialogue**

| | |
|---|---|
| **Policy analysis** | - by experts and stakeholders<br>- possible policy responses<br>- impacts under each risk scenario |
| **Policy discussion** | - by experts and stakeholders<br>- possible policy responses<br>- impacts under each risk scenario |
| **Policy advice** | - POLICY REPORT<br>- discussion with government |

The framework proposed here outlines a method for structuring a complex analytical task when there are uncertainties about the very nature and scope of the threats to which a policy response is needed. It is a tool for conceptualising and managing risk and uncertainty in the context of transitory food insecurity policies. It also provides information on how the interactions between policies and the different scenarios can play a role in policy design and the likely consequences in term of policy and diversification. The development and refinement of this framework has benefited from the experience of its successful application to the study on transitory food insecurity in Indonesia. However, it should be stressed that the application of the framework will always need to be adapted to the reality of data and stakeholders in each country.

## *Notes*

1. The concept of food security, its measurement and trends have been analysed in depth in OECD (2013a). This study refers only to the conceptual elements that are necessary to illustrate the proposed methodology.
2. See also Schmidhuber and Tobiello (2007).
3. http://www.who.int/nutgrowthdb/about/introduction/en/.
4. See Box 3.1.
5. The Communication from the European Commission on CAP reform in November 2010 establishes food security as the first of its three strategic aims (EC, 2010) in the preparation of the legislative proposals for the CAP post 2013. Japan defines securing stable food supply as one of the major policy goals in the Basic Law on Food, Agriculture and Rural Areas and set food self-sufficiency rate target of 50% by 2020 on calorie supply base. Similar objectives are expressed in different words in the Norwegian White Paper 2011-12, the Mexican Sectoral Programme 2007-12 and in other OECD countries like Korea and Israel. The 2012 Food Law in Indonesia establishes food security as one of its three principles. Similar provisions exist in other emerging economies as China, India and Indonesia.
6. DEFRA (2009) and DEFRA (2010) are good examples of the assessment of threats to food security in an OECD country.
7. Where empirical information is not available, the FAO method for measuring undernourishment using per capita energy consumption recommends the assumption of a log-normal distribution.
8. The distribution is of a Beta type, given it represents a possible range of occurrences of prevalence of food security that may take values only between 0 and 1, and is expected to be very skewed to the left.
9. If food production information is available in the survey it would be useful to use it to analyse impacts of production shocks. The SUSENAS database includes own-consumption and this information can also be used to analyse production and income shocks.

# Annex 2.A

## A formal model for transitory food insecurity policy design

This Annex presents the optimisation model for transitory food insecurity policy design and obtains the conditions that characterise the optimal policy strategy. The idea is that the government has some aversion to the risk of food insecurity. Because the scenarios are designed to capture only situations that negatively affect food security and a reference scenario, this implies the objective of reducing or eliminating any risks that imply lowering the of food security.

It is assumed that there are

- *n* states of nature (scenarios): 1,2... *i*,... *n*. Their respective probabilities are $\rho_i$, or $\boldsymbol{\rho}$ in vector form.
- *m* policy instruments with policy efforts designated by: $P_1$, $P_2$, ... $P_j$, ... $P_m$, or $\boldsymbol{P}$ in vector form. We assume *n*>*m*.
- Budgetary costs associated with each of these efforts contained in are a set of functions: $B_j(P_j)$.

Further assumptions are:

- $S_i$ is the prevalence of food security in state *i*. It is a function of the vector of policy efforts: $S_i = F_i(\boldsymbol{P})$, or $\boldsymbol{F}(\boldsymbol{P})$ in vector form.
- The government is averse to the risk of low levels of food security prevalence. This is reflected by a standard expected utility function E[U(***S***)], with $U' > 0, U'' < 0$.
- A maximum budget of B accounts for the opportunity costs of policy measures and constrains decisions about instrument choice.

Then, the objective of the government can be written as:

$$\max_{P} E[U(\boldsymbol{S})] \qquad s.t. \qquad \boldsymbol{S} = \boldsymbol{F}(P) \;\; and \;\; \bar{B} = \sum_{j=1}^{m} B_j(P_j)$$

The maximisation problem in Lagrangian terms takes the following form:

$$\max_{S_i, P_j, \lambda_i, \gamma} L = \sum_{i=1}^{n} \rho_i * U(S_i) \; + \; \sum_{i=1}^{n} \lambda_i * [S_i - f_i(\boldsymbol{P})] \; + \; \gamma * \left[ B - \sum_{j=1}^{m} B_j(P_j) \right]$$

The (2**n*+*m*+1) first-order conditions for a maximum are:

$$\rho_i * U'(S_i) + \lambda_i = 0 \qquad i = 1 \ldots n,$$

$$\sum_{i=1}^{n} \lambda_i * f'_{ij}(\boldsymbol{P}) \; - \gamma * B'_j(P_j) = 0 \qquad j = 1 \ldots m,$$

$$S_i = f_i(\boldsymbol{P}) \qquad i = 1 \ldots n,$$

$$B - \sum_{j=1}^{m} B_j(P_j) = 0,$$

where $f'_{ij} = \frac{\partial f_i}{\partial P_j}$.

This can be written as the condition that characterises the optimal policy strategy:

$$\frac{1}{B'_j(P_j)} * \sum_{i=1}^{n} \rho_i * U'(S_i) * f'_{ij}(P) = \frac{1}{B'_1(P_1)} * \sum_{i=1}^{n} \rho_i * U'(S_i) * f'_{i1}(P),$$

$$j = 2, \ldots, m$$

The final set of optimality conditions states that the marginal gains from different policies have to be equalised. In particular, they must be equal to the marginal gains from the reference set of policies, denoted by $P_j$. The gains from implementing a policy are expressed as cost-effectiveness ratios; that is, the marginal gain of policy *j* per dollar of additional budgetary expenditure $B'_j$ must be equal across all policy options. Gains from each policy in each state *i*, $f'_{ij}$, include their marginal impact on improving (or reducing if negative) food security prevalence measured in utility terms U' (to account for risk aversion), and weighted by the probability that this state or scenario prevails ($\rho_i$).

Figure 2.A1 illustrates with an example of two policies and two scenarios with *ceteris paribus* assumptions on other scenarios and policies. It is assumed for simplicity of graphical visualisation that the two scenarios are equally likely. The frontier S depicts the maximum level of food security prevalence that can be obtained in scenario 2, for each potential level of food security prevalence that could occur in scenario 1, given the budget constraint and the policies selected in each of the two scenarios. The graph presents the equilibrium condition under the additional assumption (for graphical convenience) of a linear budgetary constraint. At point A, the marginal gain on the prevalence of food security from policy 2 is greater than from policy 1, and therefore a shift of resources towards policy 2 is worthwhile given that it is more effective in improving food security under Scenario 2. This implies a movement from point A to point B.

**Figure 2.A1. Stylised policy choice**

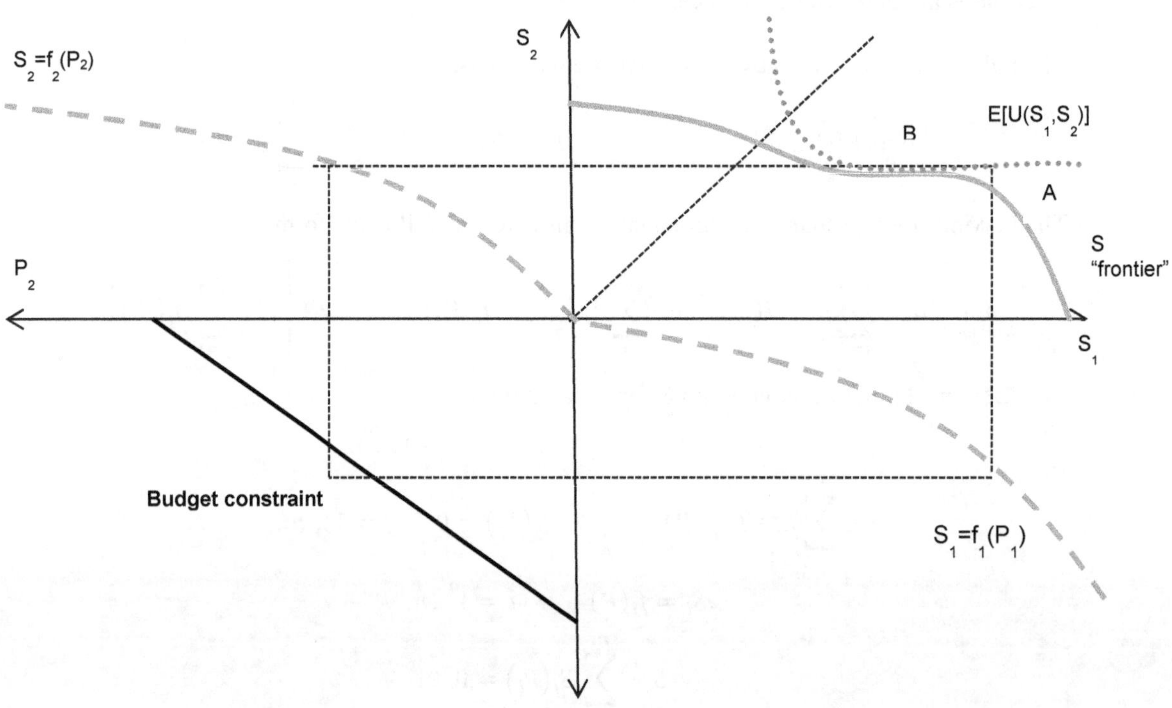

# *Chapter 3*

# Analysis of food insecurity risk in Indonesia

*The analytical framework comprises three steps: preparatory analysis, risk assessment and policy analysis. The preparatory analysis is technical in nature and involves identifying data sources, including household expenditure surveys, indicators and models. A consultation process is proposed for the second and third steps. Risk assessment draws on the perceived food security threats of experts and stakeholders and on available scientific and statistical evidence. This evidence is then transformed into a set of scenarios, each one corresponding to a specific perceived food risk, thereby allowing a rigorous assessment of its likelihood and its estimated impact on food security. The participation of experts and stakeholders is a key factor in identifying plausible food insecurity scenarios. The subsequent policy analysis focuses on existing and potentially new policy instruments, and their impacts on each scenario. A portfolio approach is then adopted to analyse policies and scenarios jointly*

## 3.1. Preparatory work (Step 1)

The Framework for the Analysis of Transitory Food Insecurity described in Chapter 2 was applied to Indonesia. This exercise involved intensive exchanges with Indonesian policy makers, experts and stakeholders following the three-step process. Three OECD missions to Indonesia were associated with these steps: a preparatory mission in June 2013, a risk assessment mission in October 2013, and a policy consultation mission in February 2014. A regional seminar in Bogor in November 2014 was an opportunity for further policy discussion of the results of the study.

Step 1 (preparatory work) led to the identification of the relevant data sources in close collaboration with the Indonesian Central Statistics Agency (BPS) and experts in the Ministry of Agriculture and in academia. The main statistical sources identified include the National Socioeconomic Survey SUSENAS, in particular its consumption module, the Consumer Price Indices database and the agricultural commodities balance sheets. The indicators available for measuring food insecurity in Indonesia were examined, including poverty and food poverty from BPS, the Food Security and Vulnerability Atlas from the World Food Programme (WFP), and the Food Security Indicators of FAO. Finally, the SUSENAS panel consumption data were used to estimate a food demand system to be used for simulating food insecurity risks. To our knowledge, this is the first estimated demand system available for Indonesia.

This chapter analyses the food security situation of Indonesia based on a variety of data sources assembled during Step 1. First, section 3.1 looks at Indonesia's evolving food security status in a regional context, while section 3.2 provides more details of food production and consumption within the country for a benchmark year, 2010. Two main sources of data are used:[1] the production balance sheets and the food security indicators of FAO (2013) and the SUSENAS data base. SUSENAS is a nation-wide household survey started in 1963-64, representing 33 provinces and seven groups of Indonesian islands (Sumatra, Jawa, Nusa Tenggara, Kalimantan, Sulawesi, Maluku, and Papua).[2] In 2007, the sample size of the panel survey was expanded to cover around 66 000 households, comprising about 260 000 individuals. It contains a core questionnaire, which collects household characteristics (location of residence, the sex, age, marital status, and education and employment of all household members, and receipt of social security or subsidised rice programme) and a consumption module questionnaire, which gathers households' consumption information. Every three years a wider SUSENAS survey is undertaken to obtain additional information covering topics such as health, education and housing. The panel part of the survey is repeated on the same households, allowing the construction of a panel data set for certain periods (such as the period 2008-10, which has been extensively used in this study).

*Overview of Indonesia's food security situation*

Indonesia is a net exporter of agro-food products but a net importer of most of its staple foods, including rice, wheat and maize. This leaves Indonesian policy-makers and consumers uneasy about the country's vulnerability to international markets for these commodities and also uneasy about unforeseen events that may impact negatively on the country's domestic food production capability. Transitory food insecurity is a concern for Indonesia not just because it is a net food importer, but also because possible international fuel price increases, macroeconomic shocks, plant and animal disease outbreaks and natural disasters all threaten the reliability of its food supplies.

While these concerns are important, recent data from the Food and Agriculture Organisation of the United Nations (FAO) indicate that food security in Indonesia has improved significantly in recent decades. Over the three-year period 1990-1992, the average prevalence of undernourishment in Indonesia was 22.2% of the population, but by the 2011-13 period this measure had declined to 9.1% (FAO, 2013).

**Box 3.1. The threshold of undernourishment**

The FAO sets the official threshold for Indonesia at 1 780 kcal. The FAO applies this threshold to a calibrated theoretical distribution in order to estimate the rate of undernourishment. The shape of the distribution is assumed to be lognormal, and it is calibrated with a mean equal to the per capita food energy availability and the variance of calorie intake from household survey data (SUSENAS). This calibration method is thought to be applicable in all countries, including those where there is not a good food consumption survey. Given that Indonesia has a good household expenditure and consumption survey SUSENAS, this policy study used the full distribution of households to estimate the rate of undernourishment rather than an assumed distribution. Directly applying the FAO threshold to the survey distribution leads to a prevalence of undernourishment that is systematically higher than the official FAO estimate. To avoid this mismatch, the threshold was recalibrated in order to reproduce a rate of undernourishment that is consistent with the official number published by the FAO and known by policy makers and stakeholders: 13% in 2008-10. This study therefore re-calibrates the undernourishment threshold at 1 370 kcal per day per capita. When applying this threshold to the SUSENAS distribution in 2010, a 13% prevalence of undernourishment is obtained. SUSENAS has the limitations of any household expenditure survey, with potential errors in the calculation of nutrient intake, in particular for readymade food that is more consumed with higher income. If such a bias exists, this could induce underestimations of income elasticities.

There are also some differences between the shares of calorie and protein intake from SUSENAS and the FAO's shares of supply. The share of soybeans in FAO's protein and calorie supplies are only about 2%, and the share of maize in energy and protein supply is, respectively, 8% and 13%, well above the shares for grains in SUSENAS. These differences are not surprising because the shares of intake and supply are quite different concepts. For instance, readymade food is correctly not included in the supply statistics, but rightly included in the SUSENAS expenditure statistics. It represents an important share of food expenditure in Indonesia: 12% among farm households and 16% of the rest.

Figure 3.1 compares Indonesia's performance with that of five other neighbouring countries in Southeast Asia for which comparable data are available in the FAO database. Although the measured prevalence of undernourishment was falling in Indonesia over this two-decade period, the average rate of decline per year was smaller than in any other Southeast Asian country for which data are available except the Philippines, another importer of staple foods. Moreover, in both Indonesia and the Philippines the prevalence of undernourishment increased in the early to mid-2000s, only to decline again in the years following 2005. In the case of Indonesia, this decline in measured undernourishment was continuous even during the global food price crisis of 2007-08, when the prices of Indonesia's main food imports doubled for most staple food commodities and trebled for some, including rice. The relationship between food security and exogenous shocks is clearly not simple.

**Figure 3.1. Rate of undernourishment, Southeast Asia, 1990-2012**

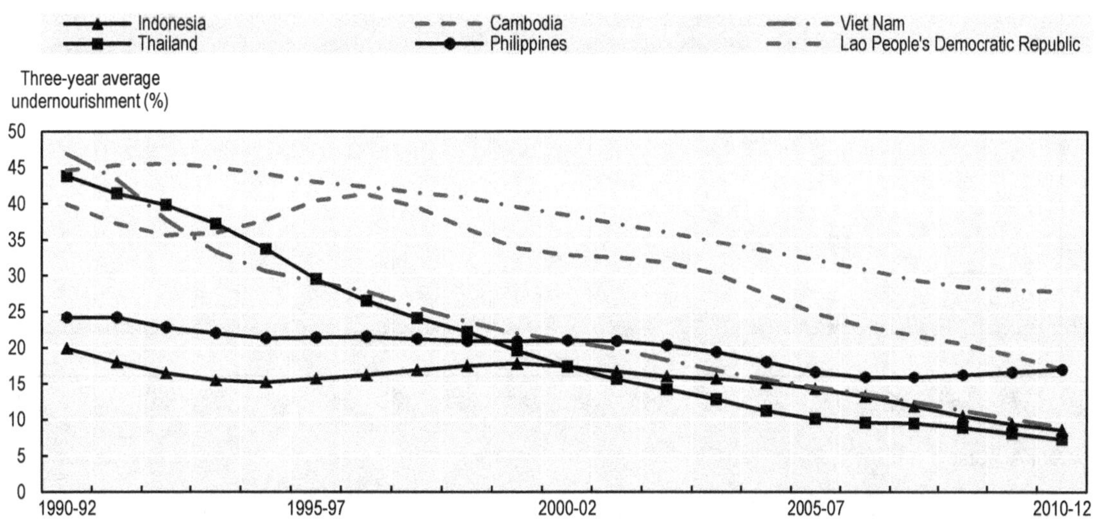


*Source*: Data from FAO *Food Security Indicators*, 2012.

According to an alternative indicator, the Global Hunger Index (IFPRI, 2013), Indonesia ranked 23rd among 78 developing countries, behind China, Malaysia, Thailand and Viet Nam, but ahead of the Philippines, Cambodia and Lao PDR. Indonesia's GHI improved from 19.7 in 1990 to 10.1 in 2013.

### *The state of food security within Indonesia*

As shown in the previous section, Indonesia has made good progress in improving its food security situation in the last two decades according to the FAO food security indicators. The prevalence of undernourishment fell from 20% of the population in 1999-2001 to 13% in 2008-2010 and 9.1% in 2011-13. Other indicators, like the depth of the food deficit, follow a similar trend. This evolution is the result of decades of high growth rates and reductions in poverty. This subsection examines these trends in more detail.

#### *Food availability*

Indonesia is a net agro-food exporting country in value terms. Food exports are mainly perennial crops such as palm oil, coffee, coconuts and cocoa beans, while non-food agricultural exports include natural rubber. Major food imports are wheat, soybeans, sugar and animal products. Rice is a staple and its domestic production has grown to more than 60 million tons in the last decade, keeping pace with domestic consumption. Rice imports have been historically below 4% of total consumption, except in occasional years (such as in the second half of the 1990s) when imports rose to above 10% (Figure 3.2). Rice imports and exports are currently controlled by a system of import licenses.

Soybeans are an important source of protein for Indonesians and imports have increased in the last two decades. Indonesia's agroclimatic conditions are not particularly suitable for growing this crop. Domestic production of soybeans fell by half, from a peak of more than 1.5 million tons during the 1990s, and has since remained at around 0.7 million tons, less than 50% of domestic consumption (see Figure 3.3).

**Figure 3.2. Rice production and trade flows, 1961-2012**

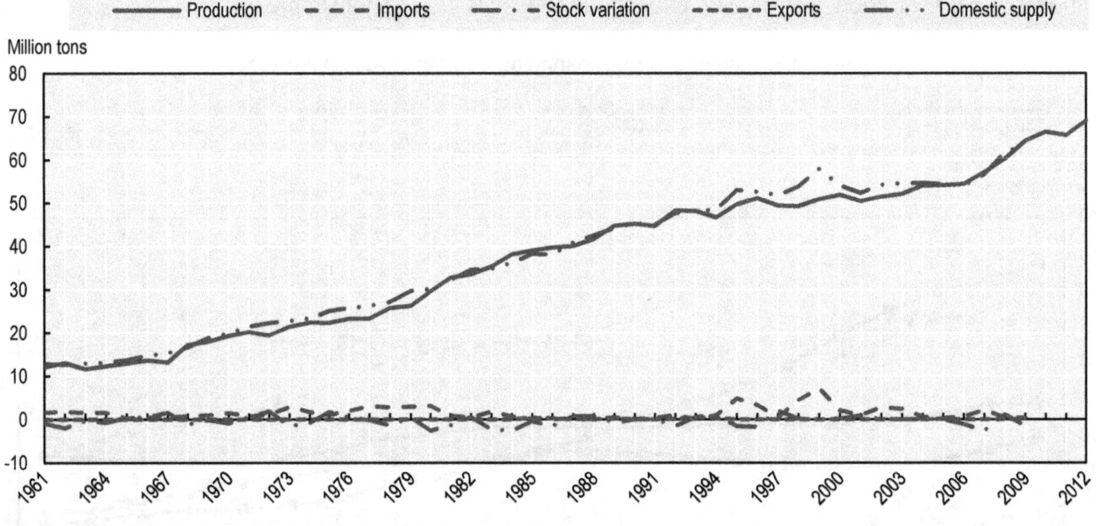


*Source*: FAO Stat.

**Figure 3.3. Production and trade of soybeans in Indonesia, 1961-2012.**

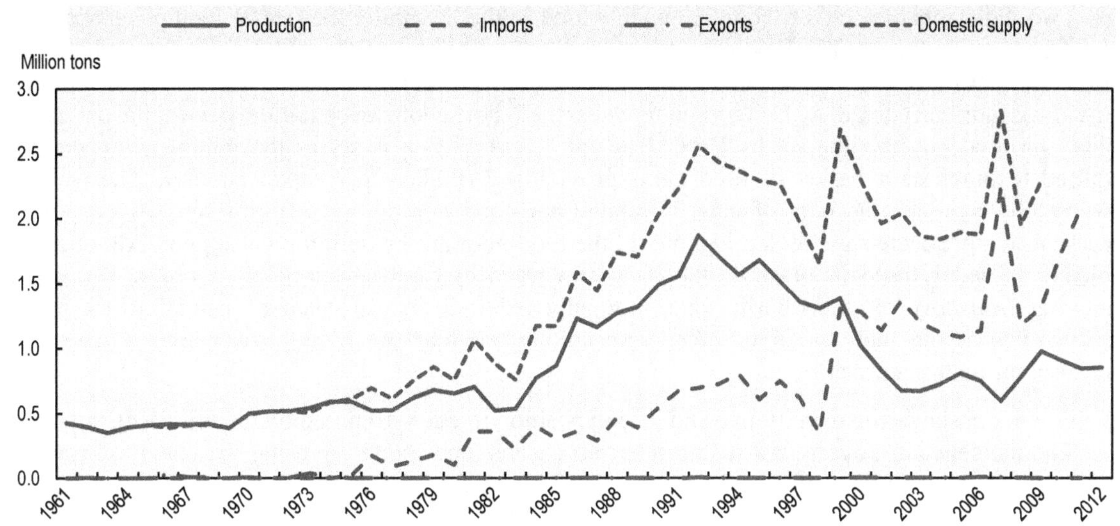

*Source*: FAO Stat.

According to Indonesia's 2010-14 Agriculture Strategic Plan, the government follows attentively the production and self-sufficiency performance of five main food commodities: rice, soybeans, sugar, maize and beef. Maize is mainly used for animal feed in Indonesia and its production has significantly increased in recent decades following the pace of demand. The production of sugar and beef has stagnated in the last two decades and imports have been growing to meet the demand. Figure 3.4 presents a summary of these trends.

**Figure 3.4. Production index of five main food commodities (1960 = 100), 1961-2012**

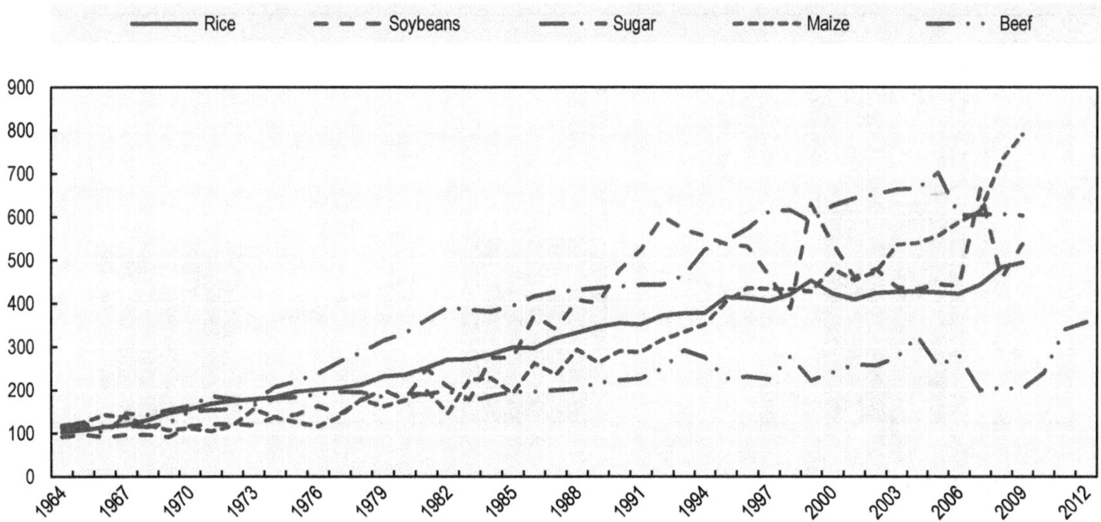


*Source*: FAO Stat.

*Access to food: Poverty and prevalence of undernourishment*[3]

Two indicators of access to food are widely used in the literature: the rate of food poverty and the rate of undernourishment. The food poverty rate indicates the share of the population whose purchasing power is insufficient to buy enough food to maintain a minimum sustainable level of calories. Individuals are considered to be food-poor when their household expenditure per capita per day is below the food poverty line set by BPS.[4] The food poverty line is set as the minimum expenditure required to purchase a basket of food items providing 2 100 kcal per capita per day. The (overall) poverty line also takes account of other minimum non-food expenditures. In this sense, food poverty represents a more acute state of deprivation: all the food-poor are by definition also poor. Although this indicator is widely used to illustrate the long-term trend of food insecurity, it is not well-suited to capturing *transitory* food insecurity since normally external shocks, such as natural disasters, are accompanied by an increase in non-food expenditure, so that the food poverty rate is seemingly increased on such an occasion.

Figure 3.5 shows the poverty rate and its composition in 2008-10 based on the official BPS poverty line. The incidence of poverty is decomposed into chronic and transitory using the panel information. The conventional definition of chronic poverty denotes households or individuals that are identified as poor in five consecutive years. Because of data constraints, the definition used here is based on a shorter panel of just three years and denotes as chronically poor those households that are poor in one of the years and poor or near-poor in the other two years. The near-poor households are conventionally defined as those whose total expenditure is less 20% above the poverty line. Households in one year but which are not categorised as chronically poor are defined as transitory poor in that year. Thus, this definition of chronic poor is less restrictive than the conventional one, while our definition of transitory poor is more restrictive.

**Figure 3.5. Poverty and food poverty ratios, and proportion of the transitory poor, 2008-2010**

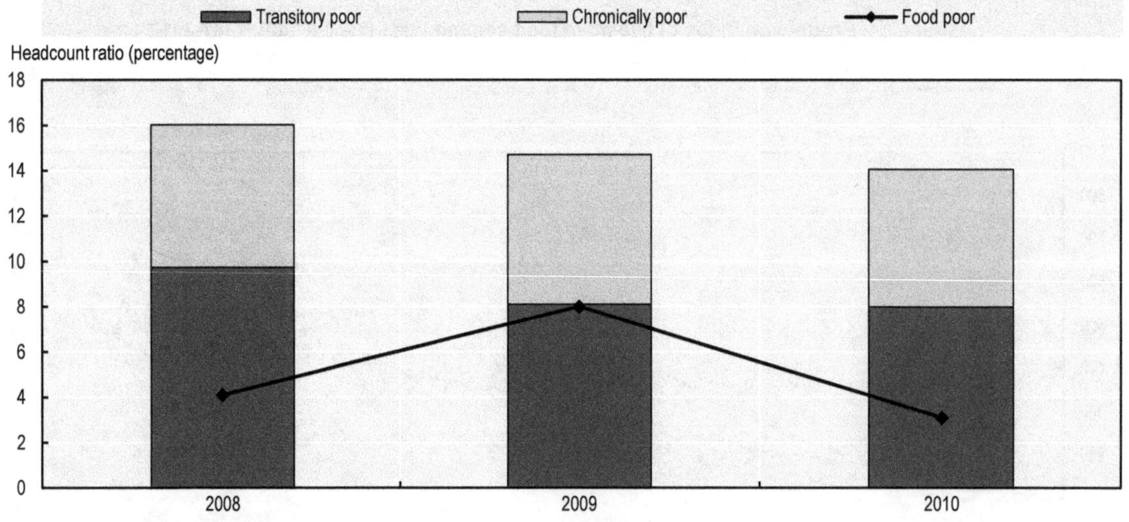


*Source*: SUSENAS Database.

Figure 3.5 shows that the poverty rate in Indonesia is around 15%, decreasing every year between 2008 and 2010. This trend is confirmed in the official rate of poverty published by the Indonesian statistical agency, which is based on the full SUSENAS sample. The data also show a much higher poverty incidence for farm households (21.5%) than non-farm households (5.3%) in 2010.[5] Within the households in the panel, 67% have been in and out of poverty during this time period. Despite the more restrictive definition used here, transitory poverty accounts for more than 60% of the poverty incidence in Indonesia in each of these three years. The reduction of the poverty rate that occurs between 2008 and 2010 comes mainly from the fall in transitory poverty.

The incidence of food poverty can also be calculated from the SUSENAS dataset and, by definition, is lower than that of poverty, in the range of 4-8% of the population. The surge in the number of food-poor in 2009 is hard to interpret and may partly be the result of a statistical effect of a re-classification of urban and rural households in that year. Food poverty is concentrated sharply in less developed regions such as Papua, Nusa Tenggara and Maluku. The incidence of food poverty among farm households (5.2%) was higher than non-farm households (2.2%) in 2010.

The second indicator of access to food is the rate of undernourishment, which is measured by household calorie consumption. An individual is defined as undernourished when daily calorie intake is less than a certain threshold. A threshold is used that replicates the FAO's estimate for 2008-2010, giving an estimated rate of undernourishment of 13% (see Box 3.1 for more details). A large share of the total population is just above the threshold and, therefore, vulnerable to becoming undernourished after an external shock. Applying to undernourishment the same definition of chronic and transitory as that applied to poverty, no more than 25% of the undernourished were chronically in that category in Indonesia in the period 2008-10 (Figure 3.6).[6] Due to the high concentration of population near the threshold, a large share of the undernourished population moves in and out of this category from year to year. The vulnerability of this group implies considerable potential instability of the rate of undernourishment as an indicator of food insecurity for the Indonesian population.

**Figure 3.6. Prevalence of undernourishment in Indonesia, chronic and transitory**

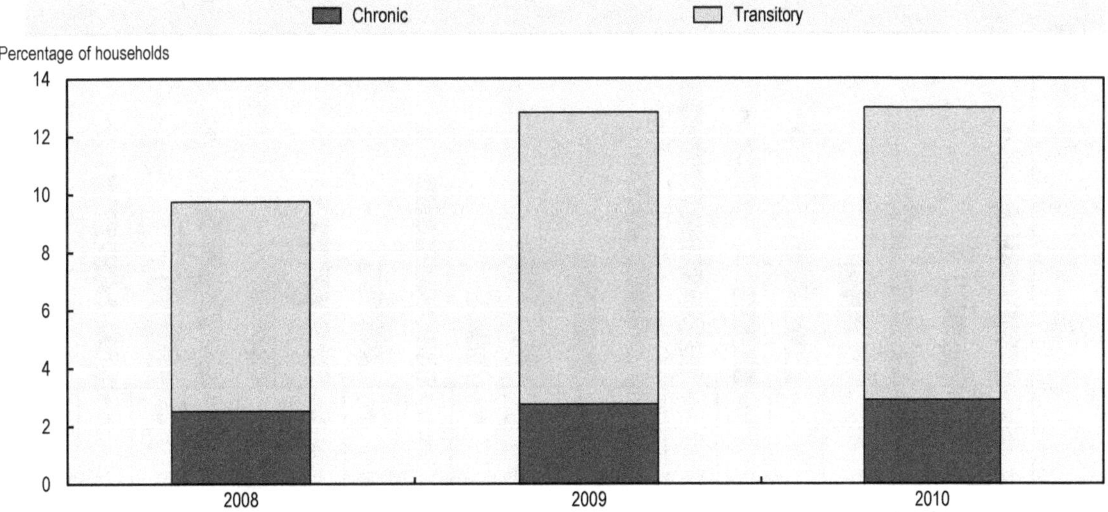


*Source*: SUSENAS Database.

The mean calorie intake for the 2010 sample is 2 030 kcal and the standard deviation is 596 kcal. The situation of undernourishment is different across islands and regions. Using the national threshold, a higher proportion of the population in Maluku and Papua (24% and 18.7%, respectively) is undernourished compared with those who live in other regions (Figure 3.7). Although farm households have a higher incidence of poverty and food poverty, the incidence of undernourishment is lower among farm households (12.1%) than among non-farm households (13.3%). Similarly, the rate of poverty is higher in rural areas (17% as compared to 10% for urban areas), but the prevalence of undernourishment is lower (12% as compared to 15% for urban). This could be explained by lower incomes in rural areas, but easier access to food through own-production or community networks that are less frequent in urban areas.

**Figure 3.7. Prevalence of undernourishment and poverty rate in 2010**

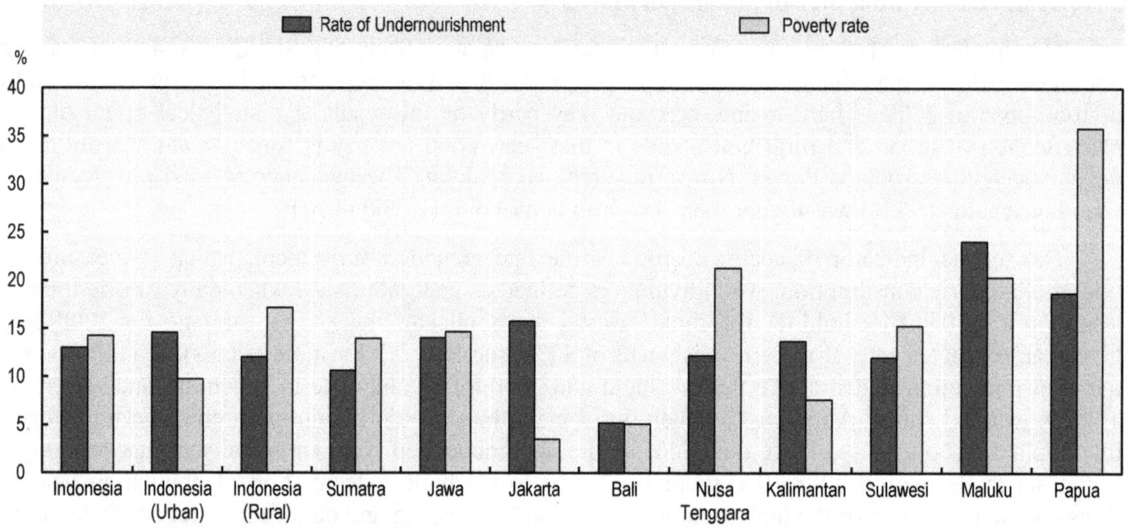

Rate of Undernourishment
Poverty rate
%
40
35
30
25
20
15
10
5
0
Indonesia
Indonesia (Urban)
Indonesia (Rural)
Sumatra
Jawa
Jakarta
Bali
Nusa Tenggara
Kalimantan
Sulawesi
Maluku
Papua

*Source*: SUSENAS Database.

**Figure 3.8. Per capita expenditure and share of expenditure by region in 2010**

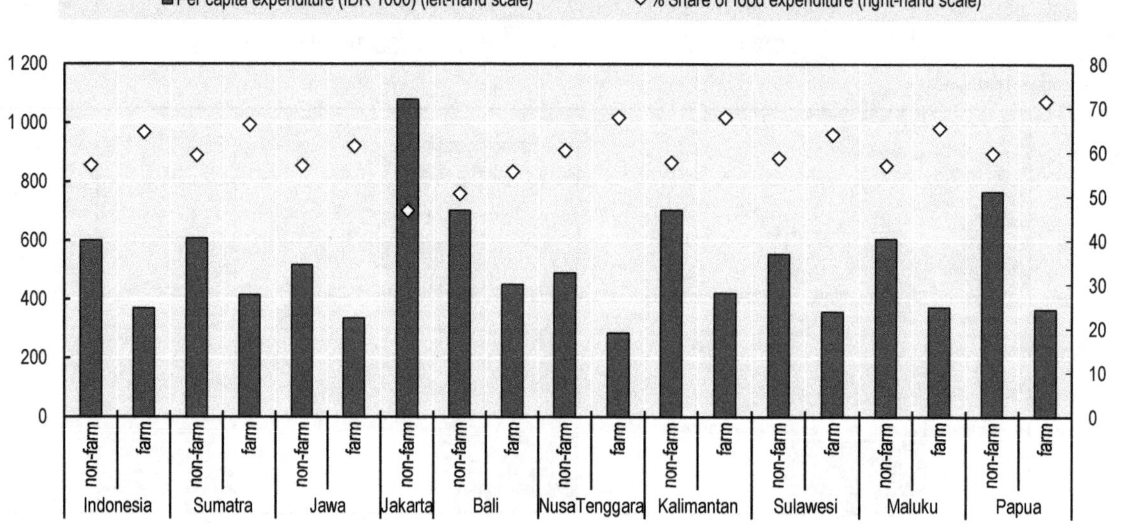

Per capita expenditure (IDR 1000) (left-hand scale)
% Share of food expenditure (right-hand scale)
1 200
1 000
800
600
400
200
0
80
70
60
50
40
30
20
10
0
non-farm
farm
Indonesia
Sumatra
Jawa
Jakarta
Bali
NusaTenggara
Kalimantan
Sulawesi
Maluku
Papua

*Source*: SUSENAS Database.

There is not a strong geographical correlation between the poverty and undernourishment indicators (Figure 3.7). For instance, Jakarta, a big city, has a very low poverty rate (below 5% as measured by expenditure and income), but it also has high rates of undernourishment – above 15%. By contrast, Papua has much higher rate of poverty than its rate of prevalence of undernourishment.

At the micro level, there is no systematic relationship between poverty and undernourishment. Only 5% of households are at the same time both poor and undernourished. The additional 8% of households suffering from undernourishment are not poor. Nine per cent of all households are poor but have sufficient energy intake from food.

The consumption module of SUSENAS shows on average around a 60% share of expenditure dedicated to food, with higher per capita expenditure and a lower share of food in expenditure among non-farm households (Figure 3.8). In line with Engel's Law, per capita expenditure is the highest in

Jakarta, while the share of food in total expenditure is the lowest (45%). At the other extreme, farm households in Papua have very low per capita expenditure and a very high share for food (more than 70%).

*Nutritional outcomes*

Even if a household has potential access to food due to its income level, consumption decisions lead to different nutritional choices.[7] The largest share of food expenditure in Indonesia (Table 3.1) is rice for farm households (16%) and, for non-farm households, readymade food (16%) followed by rice. On average, rice has by far the largest share in both calorie and protein intake in Indonesia with 44% and 50% of household calorie intake for non-farm and farm households, respectively.[8] On the other hand, rice accounts for a smaller share of food expenditure. Seafood and soybeans/nuts are important sources of protein intake accounting for around 15% and 8% in protein intake, respectively.

**Table 3.1. Composition of food expenditure, calorie intake and protein intake in 2010**

| | **Share in expenditure, %** | | **Share in calorie intake, %** | | **Share in protein intake, %** | |
|---|---|---|---|---|---|---|
| | Nonfarm | Farm | Nonfarm | Farm | Nonfarm | Farm |
| Grains/tubers | 0.9 | 2.7 | 2.8 | 6.8 | 1.7 | 4.2 |
| Rice | 10.6 | 15.9 | 44.1 | 50.1 | 36.9 | 44.4 |
| Seafood | 4.9 | 6.1 | 2.5 | 2.6 | 13.9 | 15.1 |
| Meat | 1.7 | 1.7 | 2.1 | 1.3 | 4.3 | 2.7 |
| Eggs/milk | 3.0 | 2.4 | 3.1 | 1.8 | 6.0 | 3.9 |
| Vegetables | 4.7 | 6.5 | 2.0 | 2.5 | 4.5 | 6.8 |
| Soybeans/nuts | 1.8 | 1.9 | 2.8 | 2.5 | 8.8 | 7.9 |
| Fruits | 2.5 | 2.8 | 2.1 | 2.4 | 0.9 | 1.0 |
| Oil/fats | 2.3 | 3.1 | 12.2 | 12.8 | 0.6 | 1.1 |
| Ready-made food | 16.4 | 11.9 | 22.4 | 14.1 | 18.9 | 9.9 |
| Seasoning | 1.3 | 1.6 | 0.8 | 0.8 | 1.2 | 1.2 |
| Other food | 1.4 | 1.5 | 3.1 | 2.3 | 2.3 | 1.8 |
| Non-food | 48.4 | 41.8 | - | - | - | - |

*Source*: SUSENAS Database.

**Table 3.2. Composition of calorie intake by region**

| | **Sumatra** | | **Jawa** | | **Jakarta** | | **Bali** | | **Tenggara** | | **Kalimantan** | | **Sulawesi** | | **Maluku** | | **Papua** | |
|---|---|---|---|---|---|---|---|---|---|---|---|---|---|---|---|---|---|---|
| | Non-Poor | Poor | Non-Poor | Poor | Non-Poor | Poor | Non-Poor | Poor | Non-Poor | Poor | Non-Poor | Poor | Non-Poor | Poor | Non-Poor | Poor | Non-Poor | Poor |
| Gains/tubers | 2 | 2 | 3 | 7 | 2 | 1 | 2 | 4 | 7 | 13 | 3 | 2 | 5 | 10 | 13 | 20 | 15 | 50 |
| Rice | 48 | 61 | 44 | 55 | 36 | 52 | 49 | 64 | 55 | 60 | 46 | 61 | 49 | 57 | 38 | 37 | 36 | 22 |
| Seafood | 3 | 3 | 2 | 1 | 2 | 1 | 2 | 1 | 2 | 1 | 4 | 3 | 4 | 3 | 5 | 4 | 4 | 2 |
| Meat | 2 | 1 | 2 | 0 | 4 | 1 | 4 | 1 | 2 | 0 | 3 | 1 | 1 | 0 | 1 | 1 | 4 | 2 |
| Egg/milk | 3 | 2 | 3 | 1 | 5 | 3 | 3 | 1 | 2 | 0 | 4 | 2 | 3 | 1 | 2 | 1 | 4 | 1 |
| Vegetables | 2 | 2 | 2 | 2 | 1 | 1 | 2 | 2 | 3 | 3 | 2 | 2 | 2 | 2 | 2 | 3 | 3 | 3 |
| Soybeans/nuts | 2 | 1 | 4 | 3 | 3 | 3 | 3 | 2 | 3 | 2 | 2 | 1 | 2 | 1 | 1 | 0 | 3 | 3 |
| Fruits | 2 | 2 | 2 | 1 | 2 | 1 | 3 | 2 | 2 | 2 | 2 | 1 | 4 | 3 | 5 | 4 | 3 | 2 |
| Oil/fat | 15 | 13 | 12 | 11 | 12 | 11 | 10 | 8 | 9 | 8 | 12 | 11 | 12 | 11 | 18 | 18 | 14 | 10 |
| Ready-made/drinks | 17 | 11 | 23 | 14 | 28 | 21 | 20 | 13 | 15 | 10 | 19 | 13 | 16 | 10 | 13 | 10 | 11 | 4 |
| Seasoning | 1 | 1 | 1 | 1 | 1 | 1 | 1 | 0 | 1 | 0 | 1 | 1 | 1 | 1 | 1 | 1 | 1 | 0 |
| Other foods | 3 | 2 | 3 | 2 | 4 | 4 | 2 | 1 | 2 | 1 | 4 | 3 | 2 | 1 | 2 | 2 | 3 | 1 |

*Source*: SUSENAS Database.

The share of calorie intake from different types of food is different across regions (Table 3.2). Non-poor households in Jakarta derive 28% of their dietary energy intake from readymade food. At the other extreme, the poor in Maluku or Papua consume a lot of tubers, which represent, respectively, 20% or 50% of daily calorie intake.

### *Modelling tools used for the analysis*

As well as the availability of relevant data sources, Step 1 of the analytical process involves identifying the presence of modelling tools that could be used for the analysis. Two models play a crucial role in generating the results obtained in Steps 2 and 3. First, a complete empirical household demand system was estimated specifically for this study, using panel data obtained through SUSENAS. The details of this model are given in Annex 3A. The econometric estimation of the demand system yielded elasticities for each of the 60 000 households, which are shown below.

Figure 3.9 shows the average own-price elasticities of some selected foods by quintile class of household expenditure. The own-price elasticity of rice is rather small, below 0.5 in all five expenditure groups. This means that, in the case of a spike in the rice price, households will reduce rice consumption but by less than needed to keep rice expenditure constant, so that the share of expenditure going on this staple food increases. On the contrary, the own-price elasticity of meat is very high across all income classes. Other product elasticities tend to lie between these extreme values.

The price elasticity of rice increases with income level: a 10% increase in the rice price causes a 4% fall in demand among poor households, but only 0.3% fall in the richest 20% of households.

**Figure 3.9. Elasticities of demand with respect to price, by expenditure level**

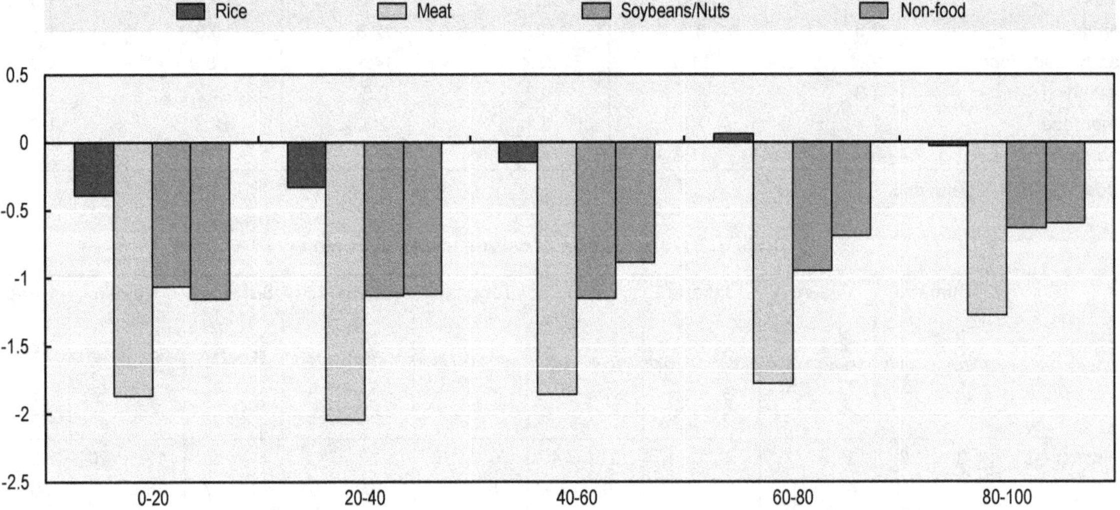


Figure 3.10 presents the expenditure elasticities obtained from this model. In general, the higher the expenditure class, the lower the income elasticity of food demand. Non-food commodities and meats are, as expected, "luxury" or superior goods for the lowest expenditure class: their demand increases more than proportionally with total expenditure (i.e. their expenditure elasticities are greater than one). Expenditure elasticities of meat exceed 0.8 for all quintiles except for the 20% households with the highest income. On the other hand, the demand for rice is much less elastic to income change than other food and non-food commodities. Indeed, rice becomes an inferior good for the richest 60% of households (negative average elasticity).

A second modelling tool plays an important role in the analysis, namely the general equilibrium model for Indonesia known as INDONESIA-E3 (Warr and Yusuf, 2014). This model is used to simulate the price changes for different commodities that would results from the shocks implied by each scenarios, and also the way these price changes are modified by different policy approaches. Thus, this model forms the link between the stylised scenarios, and the modelled demand system. Another important use of INDONESIA-E3 in the study is to ensure the quantitative consistency of each scenario[9]. An overview of this model and its use is provided in Annex 3B.

**Figure 3.10. Elasticities of demand with respect to expenditure, by expenditure level**

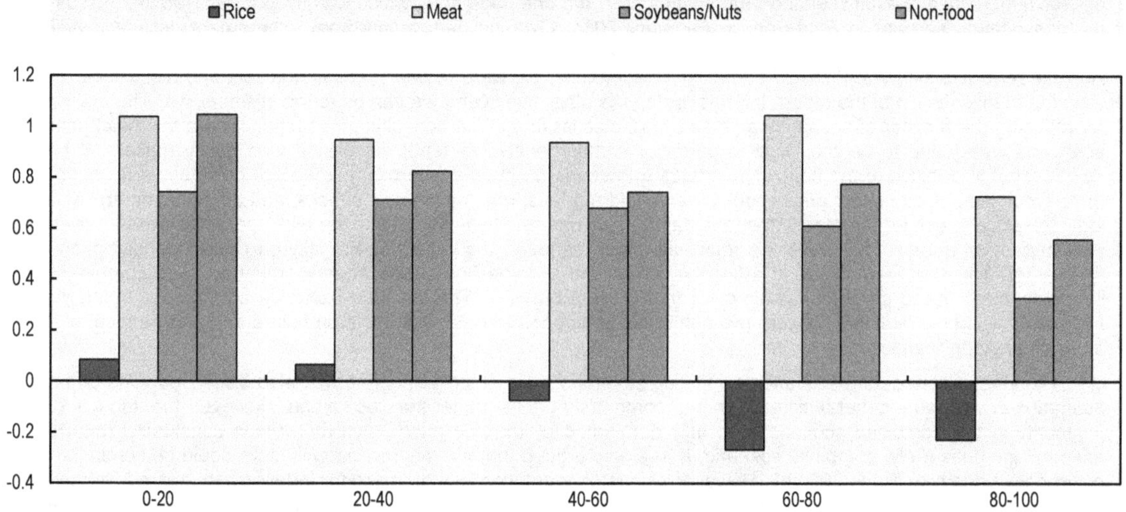


## 3.2 Analysis of food insecurity risk in Indonesia: Risk assessment (Step 2)

The second step, Risk Assessment, began by gathering information from experts and stakeholders to identify the risks of food insecurity in Indonesia. Both interviews and an electronic questionnaire were used. The initial responses on food insecurity risks were compared with available studies and the assessment of experts. In order to quantify the various food insecurity scenarios, experts on Indonesia prepared a background report with quantitative information on the likelihood of each scenario and its most likely impacts on food insecurity as measured by the prevalence of undernourishment,[10] and the material is reproduced in this section of the current report. The tools used for quantifying the scenarios include the two models described in Annexes 3A and 3B.

The assessment of food insecurity risks in Indonesia was undertaken following the sequence defined in the Framework for the Analysis of Transitory Food Insecurity (section 2) and described in Box 3.2.

**Box 3.2. The risk assessment process**

Risk assessment is one of the crucial steps in the Framework for the Analysis of Transitory Food Insecurity. The objective is not to develop an exhaustive list of risks and scenarios, but to have a short list that is recognised as relevant and plausible by stakeholders and experts. This is why respecting a consultative process for identifying the scenarios is as important as the scenarios that are finally retained.

*Risk identification.* In September 2013 an electronic survey was carried out among policy makers and experts to gather information on their perceptions of the risks of food insecurity. In October 2013, two roundtables gathered stakeholders and experts in Indonesia in order to identify the perceived risks and threats to the food security. The first roundtable took place in the Ministry of Agriculture in Jakarta and assembled policy makers and officials from different units, ministries and agencies. The second roundtable was held in Bogor with the participation of experts from academia and international organisations. As a result of the survey and the discussions, a set of 11 risk scenarios was identified for further analysis and evaluation. Table 4.1 lists and describes all 11 scenarios.

*continued*

*Risk analysis.* Existing empirical literature and data were used as the main sources for the scenario definitions. The 11 scenarios were then quantified using two modelling approaches: the general equilibrium model INDONESIA-E3 was used to ensure the quantitative consistency of each scenario and to translate external shocks into market price changes, and the LA-AIDS demand system, estimated using the SUSENAS database 2008-10, was used to simulate all the survey households' consumption responses to the price changes induced by these external shocks.

*Risk evaluation.* The demand system simulates not only how consumption expenditure reacts at micro-level in each risk scenario, but also the implications for the nutritional status of all the survey households, the rate of undernourishment among urban and rural households, the depth of undernourishment, protein intake and regional impacts. The risk assessment results of each scenario were summarised in a one page note, which was used to facilitate the discussion at the consultation seminar in Bogor on 26 February 2014. Over 60 participants from government, including high-level officials, international organisations (WB, WFP, UN-CAPSA), research centres (ICASEPS) and universities (Bogor Agricultural University) attended. Of the initial 11 scenarios, six were retained for full analysis and discussion, and are reported in this section of the report. Information on the other five scenarios can be found in Annex 4A. The main reason for reducing the number of scenarios emerged from discussions in the consultation seminar, where the initial number of scenarios was found to be too large to permit a comprehensive in-depth discussion with policy makers. It became obvious that in addition to the benchmark status quo scenario, at least one example should be retained for each of the three typologies of scenarios: price shocks from world markets, macroeconomic shocks, and domestic natural disasters. Only the price shock for the most important staple, rice, was retained. Soybean price hike scenarios were analysed, but their impact on undernourishment was relatively minor, although the impact specifically on protein consumption was a little larger. The scenarios depicting a financial crisis and a domestic macroeconomic downturn were combined into a single scenario based on an economic crisis that occurred in1997. The scenario exploring an increase in international fuel prices was also retained, as were two natural disaster scenarios: a systemic crop failure and a more local event like an earthquake in Sumatra.

The idea of considering scenarios combining several shocks at the same time was also discussed. One of the initial scenarios envisaged a general increase in four commodity prices, rather than rice alone. However, the impacts on food insecurity in this combined scenario were fully dominated by the rice price, such that it was considered redundant to examine the (less likely) combined scenario. It was also argued that the macroeconomic crisis could be combined with a commodity price hike like in 2007-08. However, this later event was less dramatic for Indonesia compared with the 1997 financial crisis, which was independent of a commodity price shock. All these reasons underlie the choice of the six scenarios that were retained.

## *Risk identification and analysis*[11]

**Table 3.3. Scenarios defined by the risk identification and quantification process**

| Scenarios retained for full analysis | Scenarios identified but not retained |
|---|---|
| **0: Reference period without any shock**<br>Probability: Once in 2 years | |
| **World market price shocks** | |
| **1: Rice price spike in international markets**<br>International price of rice ↑ by 100%<br>Probability: Once in 30 years | **6. Increase in international soybean prices**<br>International price of soybeans ↑ by 118%<br>Probability: Once in 30 years |
| | **7: Increase in international prices of rice, soybeans, wheat, and maize**<br>Rice ↑ by 212%, soybeans ↑ by 118%, wheat ↑ by 184%, maize ↑ by 124%<br>Probability: Once in 40 years |
| **Macroeconomic shocks** | |
| **2: Macroeconomic crisis**<br>Factor supplies ↓ by 11.4% (except rural land & capital)<br>Probability: Once in 25 years | **8: International financial crisis**<br>Investment demand ↓ by 40%<br>Probability: Once in 30 years |
| **3: An increase in international price of fuel**<br>International price of fuel ↑ by 114%<br>Probability: Once in 20 years | |

**Table 3.3. Scenarios defined by the risk identification and quantification process (*cont.*)**

| Domestic natural disasters | |
|---|---|
| **4: Crop failure due to insect or plant disease infestation**<br>Factor productivity of paddy land ↓ by 12%<br>Probability: Once in 15 years | **9: Systemic drought in Java due to El Niño**<br>Agricultural productivity ↓ by10% within Java<br>Probability: Once in 7-8 years |
| **5: Earthquake and tsunami in Sumatra**<br>All factor supplies (capital, land, labour) ↓ by 10% in Sumatra<br>Probability: Once in 20 years | **10: Avian influenza epidemic**<br>Factor productivity of poultry production ↓ by 20%<br>Probability: Once in 15 years |

*Scenario 1: An increase in international rice prices*

Rice is Indonesia's staple food and the country has for many decades been a net rice importer. Nevertheless, Indonesia relies on world markets for only a modest part of its consumption – less than 2% in recent years. The international market for rice is thin, meaning that volumes traded globally are small relative to global consumption, and international rice prices are consequently notoriously volatile. Large surges occur on average once every 15 years or so, though unpredictably.

Major world price surges occurred in 1972-73, 1998-99 and most recently in 2007-08, when rice prices more than tripled. The 1972-73 food price shocks coincided with global petroleum price increases resulting from OPEC actions. Indonesia was a net exporter of petroleum at this time and the beneficial impact on Indonesia of the petroleum price increases partially cushioned the negative impact of the food price increases. Nevertheless, the food price increases were a source of alarm within Indonesia, as in other food importing countries. The 1998-99 financial crisis was a time of widespread economic distress within Indonesia. Food price increases coincided with large falls in incomes, magnifying the political concern about food security. In the succeeding years, heightened political concern about food security was reflected in an increasing level of trade policy protection for Indonesia's rice industry.

The transmission of international rice prices to domestic prices and the subsequent effect on food security depends heavily on the trade policy in place. At various times, the Indonesian government has used quantitative import controls, export restrictions, tariffs, domestic price controls and public rice storage to influence both the level of domestic rice prices relative to world prices and their volatility (Fane and Warr, 2009). Recent changes in Indonesian rice import policy have greatly altered the relationship between global rice prices and Indonesian domestic rice prices.

Until the early 2000s, Indonesia was the world's largest rice importer, but since then the lobbying power of pro-farmer political groups led first to heavy tariffs on rice imports and then, in 2004, to an official 'ban' on rice imports. Despite the official prohibition, limited quantities of imports are occasionally permitted. According to Fane and Warr (2008), by 2006 this policy had restricted imports to an average of about one fourth of their previous volume and had increased domestic rice prices relative to world prices by about 37%. At the same time as the import restrictions were introduced, exports of rice were also banned, in the sense that exports could only occur through the Indonesian government's logistics agency, BULOG. Private sector exports of rice are not permitted and BULOG was instructed by the government not to export any rice.

These changes to rice import policy meant that the 2007-08 world price increase was not transmitted to the Indonesian rice market (Timmer, 2008), as can be seen clearly from Figure 4.1.[12] In fact, Indonesia's quantitative restriction on rice imports had the effect that an increase in the international price merely reduced the rent associated with the limited volume of imports that was allowed. This effect would be expected provided the higher international price does not drive the quantity imported below the maximum permitted by the import restriction, thereby reducing the value of rents associated with the import licences to zero. This reduction in rents may have been a problem for the rich urban households who own the import licences, but it did almost nothing to the domestic market price for rice, which hardly changed.

It is likely that even if the import restrictions were not in place, but the export restrictions remained, an increase in international rice prices would have had only a moderate effect on domestic rice prices. Imported and domestically produced rice are close but not perfect substitutes and imports are only a small proportion of domestic consumption. If domestic rice could be exported to take advantage of much higher international prices the effect of the international price increase on the domestic market price would be larger.

Not all households are affected in the same way by an increase in the domestic price of rice. Households that are net sellers of rice will benefit from an increase in rice prices and this will include many, but not all rural households. Net purchasers of rice will lose and this will include many but not all urban households. That is, there are both gainers and losers from an increase in rice prices. The question is how many households belong to each of these two categories.

Figure 4.1 summarises data on net sales of rice from the Indonesian Family Life Survey (IFLS),[13] which collects data on both sales and purchases of rice at household level (unlike the SUSENAS survey, which does not collect data on sales). Negative net sales mean that the household is a net purchaser. Not surprisingly, the mode, median and mean of the distribution of urban households are all negative, indicating net purchasers predominate. At the same time, over half of all rural households are net purchasers as well. Rural households include rice farmers who are net sellers of rice, and these households benefit from an increase in the price. But rural areas also include farmers who sell other commodities and purchase rice, and households that sell their labour and purchase rice. Because net purchasers outnumber net sellers, an increase in domestic rice prices lowers food security in both urban and rural areas.

Scenario 1 was conducted in two versions. In both, the international price of rice increases by 100%, as occurred in 2007-08. This is a real price increase because all other international commodity prices are held constant. The two versions of this scenario differ according to assumptions about Indonesia's trade policies at the time, demonstrating the manner in which the impact of the external shock depends on Indonesia's policy environment at the time the shocks occur.[14]

**Figure 3.11. Indonesia: Distribution of net sales of rice of urban households, 2007**

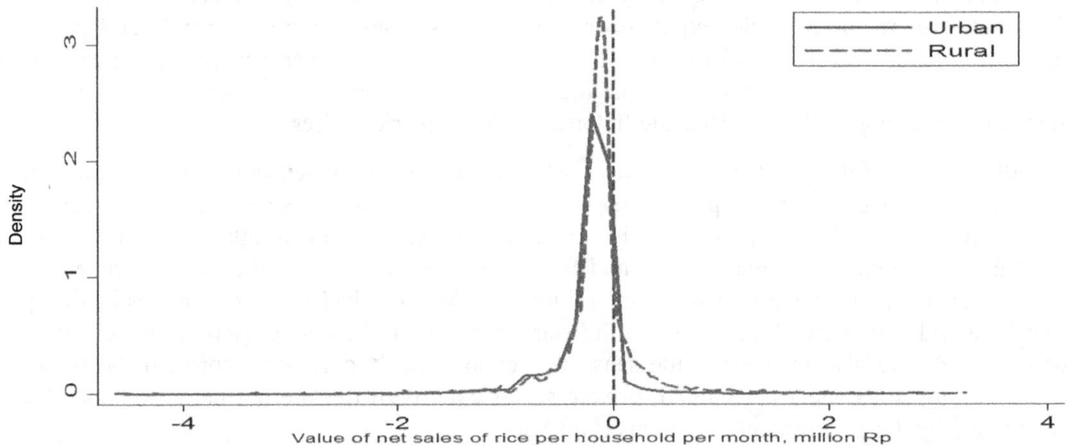


*Source*: Peter Warr's calculations from Indonesian Family Life Survey, 2007.

Scenario 1-a assumes a binding quantitative restriction on rice imports at the current level and that the volume of rice exports is held constant at zero, reflecting the official ban on rice exports. In these circumstances, transmission of t²he international price change to the domestic market is expected to be negligible, and hence domestic rice prices would barely change. It is therefore expected that food insecurity, as measured by the prevalence of undernourishment, would not be affected to any extent.

Scenario 1-b assumes that rice import policy remains as in Scenario 1-a, but that private sector exports of rice are legally permitted. This hypothetical scenario is intended to demonstrate the role of the ban on rice exports. If private sector exports were permitted, a large increase in the international price would mean that, given the lower domestic price, exports of Indonesian rice would be highly profitable. As rice exports increased, domestic prices would rise as rice became scarcer on the domestic market. It is expected that under this scenario, food insecurity, as reflected in the prevalence of undernourishment, would increase. This scenario involves an increase in the domestic price of rice of 44%[15] and a decline in real income across all households, in particular those that are net buyers of rice. Some farm households that sell rice commercially are expected to benefit from the situation, but they are a minority.

The assessment of scenario 1 is summarised in Table 4.2. It shows that when imports are controlled and exports are banned, the rate of undernourishment does not change, but that undernourishment increases (by 10 percentage points) to 23% of the population when rice exports are permitted. In the latter case, the median calorie intake is significantly reduced by 160 kcal per day per capita, while the depth of the food deficit[16] more than doubles.

**Table 3.4. Scenario 1: International rice price spike (100%)**

| | | I-a: Trade restrictions | I-b: Exports permitted |
|---|---|---|---|
| Availability | Production Imports and stocks | Rice production virtually unchanged because the domestic producer price of rice is almost unchanged. | Rural households increase paddy production by 18.1%, but 22% of domestic rice is exported. |
| | Food price | Increase in real consumer price of rice by 0.1%. | Increase in real consumer price of rice and soybean by 44.1% and 3.3%, respectively. |
| Access | Household income and assets | Household real expenditures in urban and rural areas decline by 0.05% and 0.02% on average, respectively. | Rural and urban households experience decline in real income, mainly due to higher consumer price of rice. The magnitude depends on the share of expenditure on rice. Poor households suffer largest decline in real income. |

**B. Impacts of Scenario 1: International rice price spike**

| | Rate of undernourishment (percentage) | Median calorie intake (kcal per day per capita) | Depth of food deficit (kcal) |
|---|---|---|---|
| **Before shock** | | | |
| All Indonesia | 13 | 1941 | 19 |
| Urban households | 14 | 1883 | 21 |
| Rural households | 12 | 1974 | 18 |
| **a) After rice price hike with trade restrictions** | | | |
| All Indonesia | 13 | 1936 | 20 |
| Urban households | 14 | 1880 | 22 |
| Rural households | 12 | 1969 | 18 |
| **b) After rice price hike, exports permitted** | | | |
| All Indonesia | 23 | 1780 | 45 |
| Urban households | 23 | 1771 | 45 |
| Rural households | 23 | 1785 | 45 |

At first sight, these results might suggest that an export ban protects poor households from food insecurity. However, the negative consequences of restricting exports and other restrictive trade measures are spread over time: because of these policies, the domestic price of rice has been structurally higher than the international price in almost every year since 2000, chronically affecting the rate of undernourishment in the reference scenario, whereas the price-spike scenario is assumed to occur just once in 30 years. Moreover, the simulations assume that only Indonesia applies export restrictions during the price spike, which is unrealistic. Export restrictions by many or all countries concurrently would directly exacerbate the world market price spike and its negative consequences for the poor (FAO et al., 2011). Furthermore, there are additional long-term effects of export constraints, which discourage investment and prevent producers responding to market signals.[17]

*Scenario 2: Macroeconomic crisis*

Recessions occur from time to time in Indonesia. The most severe recession in recent years was the one that accompanied the 1997-99 Asian financial crisis. In this case real, GDP per person declined by 11% (Pardede, 1999), and real consumption and investment were similarly affected (Figure 3.12). However, recessions within Indonesia are not necessarily caused by international crises. Examples are the recessions of 1981-82 and 1985, which occurred for domestic reasons. This scenario looks at a broadly based macroeconomic slowdown caused by an economy-wide collapse in production. For convenience, it is represented as a contraction in all factor supplies of 11.4%, with the exception of capital and land used in agriculture. This makes the contraction somewhat larger and more broadly based than if it were due to a financial crisis (see scenario 8, Annex 4A).

A domestic macroeconomic slowdown affects incomes as well, and not necessarily predominantly in investment-related industries as with a financial crisis. Rather, the income effects are likely to be widely distributed across industries, though agricultural incomes tend to be less sensitive to domestic macroeconomic conditions except as a result of slowdown in demand, arising from other parts of the Indonesian economy. Nonetheless, because large numbers of Indonesian households are just above the poverty or undernourishment thresholds, an economic slowdown may have significant effects on poverty incidence and undernourishment.

**Figure 3.12. Indonesia's GDP, consumption and investment growth, 1961 to 2012**

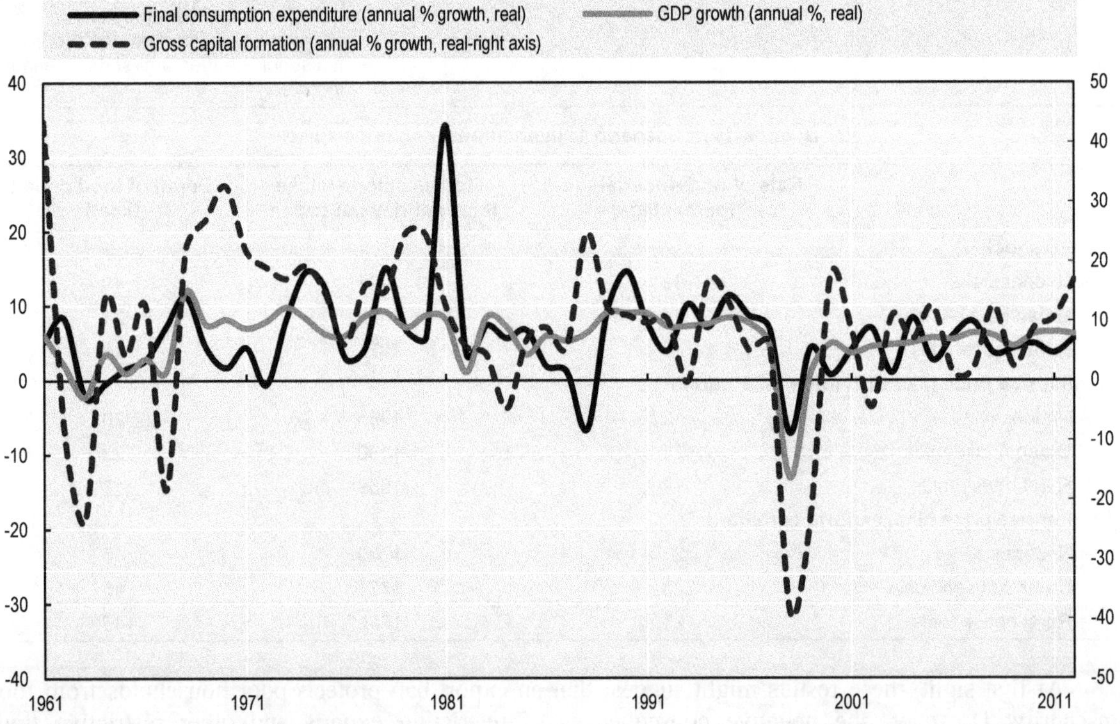


*Source*: World Bank, World Development Indicators.

**Table 3.5. Scenario 2: Macroeconomic crisis**

| | | |
|---|---|---|
| **Availability** | **Production, imports and stocks** | Production of rice and other staples remains almost constant. Imports of food grains decline by 10%. |
| **Access** | **Food prices** | Real price of rice increases by 9.8%. Real prices of staples in general increase by 8%. |
| | **Household income and assets** | Real incomes of urban and rural household decline by 15.3% and 11.2%, respectively. |

**B. Impacts of Scenario 2: Macroeconomic crisis**

| | Rate of undernourishment (percentage) | Median calorie intake (kcal per day per capita) | Depth of food deficit (kcal) |
|---|---|---|---|
| **Before the shock** | | | |
| All Indonesia | 13 | 1 941 | 19 |
| Urban households | 14 | 1 883 | 21 |
| Rural households | 12 | 1 974 | 18 |
| **After the economic shock** | | | |
| All Indonesia | 22 | 1 771 | 40 |
| Urban households | 23 | 1 731 | 42 |
| Rural households | 21 | 1 795 | 39 |

This scenario examines the effect of an 11% contraction in GDP, modelled as an 11.4% contraction in factor supplies other than agricultural land and capital used in agriculture. A significant increase in undernourishment is expected.

The results show that the loss of real incomes leads to reduced capacity to purchase goods and services, even necessities (Table 4.3). Food consumption is affected less than most other goods, but it has more severe impacts on low-income households with higher shares of food expenditure. Urban households are more affected than rural households that produce food. Real GDP per capita declines on average by 9.7%.

The macroeconomic crisis has very strong consequences for food security, with the rate of undernourishment nearly doubling. The impact is only slightly greater in urban areas where the fall in economic activity is stronger. However, the reduction in the median calorie intake is smaller in urban areas where the middle income classes seem more resilient to the macroeconomic shock.

*Scenario 3: An increase in the international price of fuel*

In recent decades, Indonesia has shifted from being a net exporter of petroleum to a net importer. At the end of the 1970s petroleum exports accounted for a large proportion of government revenue, through taxes, and also a large proportion of Indonesia's foreign exchange earnings. Indonesia was an important member of the Organisation of Petroleum Exporting Countries (OPEC). Today, the situation is very different. Petroleum imports are significant and subsidies on domestic fuel consumption absorb a high proportion of government expenditure. This makes Indonesia vulnerable to increases in international petroleum prices.

Since the early 1970s, three instances of sharp increases in international petroleum prices have occurred: 1973-74, when petroleum prices tripled, 1979-80, when they doubled again, and 2007-09. Between September 2007 and May 2009 the international price of petroleum rose by around 114% (Figure 4.3). This scenario assumes a similar increase (+114%) in the international fuel price. Since rates of fuel subsidy are assumed to remain constant as petroleum prices rise, the international price increases are transmitted to domestic petroleum product prices, and hence domestic petroleum prices also rise by 114%. In addition, the budgetary cost of the fuel subsidy rises as the international price rises. The effect of this price increase on food insecurity is estimated assuming that trade policy with respect to food remains at its present settings.

**Figure 3.13. International petroleum prices (USD per barrel), 1971-2013**

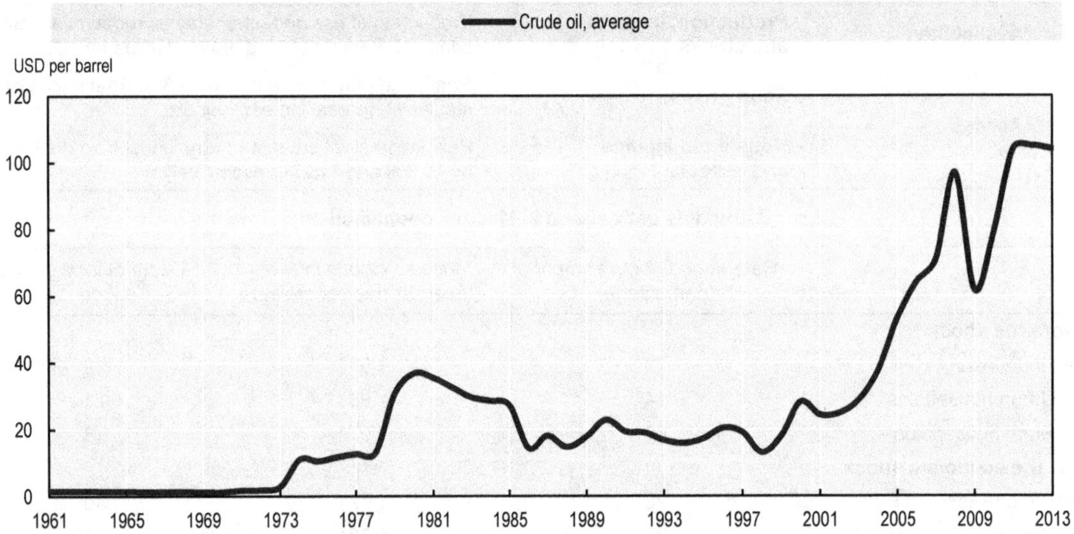

*Source*: International Monetary Fund, International Financial Statistics.

**Table 3.6. Scenario 3: Increase in fuel prices**

| | | |
|---|---|---|
| Availability | Production imports and stocks | Real costs of food production increase by 4.3 %. Total grain consumption declines slightly by 0.02%. |
| Access | Food price | Real consumer prices of rice and soybeans increase by 36.1% and 3.7%, respectively. |
| | Household income and assets | Real incomes of urban and rural households decline by 0.2% and 3.0%, respectively. |

**B. Impacts of Scenario 3: An increase in the international price of fuel**

| | Rate of undernourishment (percentage) | Median calorie intake (kcal per day per capita) | Depth of food deficit (kcal) |
|---|---|---|---|
| **Before the shock** | | | |
| All Indonesia | 13 | 1 941 | 19 |
| Urban households | 14 | 1 883 | 21 |
| Rural households | 12 | 1 974 | 18 |
| **After the fuel price hike** | | | |
| All Indonesia | 21 | 1 809 | 38 |
| Urban households | 21 | 1 791 | 39 |
| Rural households | 20 | 1 821 | 37 |

Increases in petroleum prices affect food security, first, by making Indonesia poorer in general. Because petroleum is a net import item, increases in international petroleum prices reduce Indonesia's national income by lowering its terms of trade with the rest of the world. Second, higher petroleum prices within Indonesia raise transport costs within the country for all goods, especially food, which is bulky and costly to transport, increasing domestic retail prices. Third, transport costs for people are raised, on both private and public transport, and this affects the living costs of all Indonesians. Fourth, the cost of producing food is particularly sensitive to petroleum prices because fertiliser (especially urea), pesticides, herbicides and fuel for farm machinery all depend on petroleum inputs. As the cost of petroleum rises, the cost of producing food also rises (Warr, 2011). These linkages affect food security negatively, but with time lags. Finally, petroleum prices are important direct consumption items, for both low-income household groups (kerosene) and high-income groups (gasoline).

The increase in undernourishment is slightly smaller, but the same order of magnitude, as in the previous scenarios. The prevalence of undernourishment increases by 8 percentage points to 21% of the population. The impact is a bit larger in rural areas due to the negative impacts on transportation and agricultural input prices

*Scenario 4: Crop failure due to a plant disease infestation*

Rice is grown as a monoculture and is therefore especially susceptible to disease and pest infestations. Pests such as the brown planthopper (BPH),[18] known locally as *wereng cokolat*, have been a serious problem in the past. Major outbreaks occurred during the 1974-75 planting season and there was an especially severe outbreak in 1985-86. More recently, outbreaks occurred in 1998 and 2011.

Research by the International Rice Research Institute has shown that major BPH outbreaks are a direct consequence of excessive insecticide use (Heong et al., 1998, Pingali et al., 1997), which kills its natural biological predators (Heong et al., 2013). Recognition of this point led to a decree in 1986 banning a large range of organophosphate pesticides and gradually removing pesticide subsidies. Problems with BPH abated markedly as pesticide use declined. In the last decade, this situation has been reversed. Beginning in 2002, restrictions on pesticide imports were relaxed. FAO data indicate that between 1990 and 2012 pesticide imports, mainly from China, increased from USD 1.2 million to USD 71 million. These pesticides are marketed aggressively and misleadingly within Indonesia under a variety of attractive brand names. BPH populations have gradually increased (Fox, 2014a).

In January 2014, the Department of Plant Protection at the Institute of Agriculture in Bogor issued a press release predicting a severe BPH problem in 2014.[19] The loss of 6 million tons of paddy output was predicted, equivalent to about 4 million tons of milled rice, roughly 12% of the annual crop. Based on this prediction (which did not materialise in 2014 but is still a real threat), scenario 4 assumes a widespread reduction in rice output in Indonesia, equivalent to 12% of the crop, which is modelled as a 12% reduction in the productivity of rice production. Two variants of this scenario are analysed: scenario 4-a assumes that the current restriction on rice imports remains in place whereas scenario 4-b assumes that there is a fixed *ad valorem* tariff on rice imports, set at the rate that yields the same initial volume of imports as under Scenario 4-a. If there were a large-scale pest outbreak, the Indonesian government would most probably permit increased volumes of rice imports. The purpose of the two scenarios is to show the importance of this policy response.[20]

Insect and plant disease infection may spread across a wide area of Indonesia. Under the first scenario, contraction in rice production is expected to cause rice prices within Indonesia to increase markedly because of the inelastic demand for rice in Indonesia, worsening food insecurity although farmer incomes will increase. Rural households in infected regions are the most severely affected. Rural households in other regions gain from a higher rice price if they are net sellers of rice but lose if they are net buyers. There may also be severe impacts on urban households, particularly for low-income households. In scenario 4-b, the price impact will be muted by additional imports, leading to reduced incomes among farmers but smaller effects on food insecurity.

The results in Table 3.7 confirm these expectations. A crop failure will have a large impact on undernourishment, pushing the undernourishment rate up to 25% if imports do not rapidly find their way into Indonesia (scenario 4-a). Rural households will be the most affected because of the greater impact of the crop failure in the rural economy. This strong impact on undernourishment would not occur if trade restrictions did not impede imports (scenario 4-b). In this case, the rise in the prevalence of undernourishment would be limited to 3 percentage points.

**Table 3.7. Scenario 4: Crop failure**

| | | 4-a: Trade restrictions | 4-b: Imports permitted |
|---|---|---|---|
| Availability | Production imports and stocks | 12 % loss of rice output | 12% loss of rice output. Imports increase to supplement domestic demand |
| Access | Food price | Increase in real consumer price of rice by 48% | Increase in consumer price of rice by 16.1% |
| | Household income and assets | Rural and urban households experience 1.2% and 2% decrease in real income on average, respectively. | Household income remains almost constant on average |

**B. Impacts of Scenario 4: Crop failure**

| | Rate of undernourishment (percentage) | Median calorie intake (kcal per day per capita) | Depth of food deficit (kcal) |
|---|---|---|---|
| Before the shock | | | |
| All Indonesia | 13 | 1 941 | 19 |
| Urban households | 14 | 1 883 | 21 |
| Rural households | 12 | 1 974 | 18 |
| a) After the crop failure | | | |
| All Indonesia | 25 | 1 753 | 52 |
| Urban households | 24 | 1 757 | 50 |
| Rural households | 26 | 1 751 | 53 |
| b) Imports allowed after the crop failure | | | |
| All Indonesia | 16 | 1 880 | 26 |
| Urban households | 17 | 1 850 | 26 |
| Rural households | 16 | 1 901 | 25 |

*Scenario 5: Earthquake and tsunami in Sumatra*

Indonesia is located on the seam of four major tectonic plates, one of the most seismically active regions in the world. As a consequence, earthquakes occur frequently and the country experiences some of the strongest earthquakes. Since Indonesia is composed of islands and the majority of the population live in coastal areas or nearby, the country is also highly vulnerable to tsunamis caused by these earthquakes. Sumatra Island and the surrounding small islands belong to the region where important earthquakes occurred most frequently over the past decade (Table 4.6). The region also accounts for 23% of Indonesia's rice production. Clearly, the effect of an earthquake or tsunami in Sumatra on food security within those local areas directly affected will be catastrophic. Damage to transport infrastructure may also affect vulnerable people within Sumatra who are far removed from the areas affected directly.

Scenario 5 analyses the effects on Sumatra as a whole. The INDONESIA-E3 and TERM[21] models are used in conjunction. TERM is used to obtain the estimated effects in Sumatra as a whole. These effects are then used to estimate the distributional effects across households within Sumatra using INDONESIA-E3 on the assumption that the structure of effects across households within Sumatra is approximated by the structure of these effects within Indonesia as a whole, as captured within INDONESIA-E3.

According to Gignoux and Menéndez (2013), only earthquakes with an intensity level higher than 7 are likely to affect individuals, with the most important economic damage involving asset and household income losses. They estimated that an earthquake of this magnitude reduces per capita consumption expenditure by, on average, 10% for rural households and 6% for urban households in the short term.

An earthquake will affect food security is not only via the loss of income and private assets, but also damage to public infrastructure. There is a temporary loss of physical access to food and water in the affected regions, and higher food prices in local markets. Impacts are catastrophic, but more location-specific. Households in remote areas that lose transportation infrastructure will be more severely damaged. Higher food prices have a more modest impact on low-income households in non-affected regions, operating through higher food prices.

**Table 3.8. The most important earthquakes in Indonesia, 1900-2013**

| Date | Place | Magnitude (Richter scale) | Number of deaths | Number of people affected | Damage (000 USD) |
|---|---|---|---|---|---|
| 21-01-1917 | Bali | - | 15 000 | - | - |
| 12-12-1992 | Flores Region | 7.8 | 2 500 | - | - |
| 26-12-2004 | Sumatra – Andaman Islands | 9.1 | 165 708 | 532 898 | 4 451 600 |
| 27-05-2006 | Java | 6.3 | 5 778 | 3 177 923 | 3 100 000 |
| 12-09-2007 | Southern Sumatra | 8.5 | - | - | 500 000 |
| 30-09-2009 | Southern Sumatra | 7.5 | - | 2 501 798 | 2 200 000 |

*Source:* The OFDA/CRED International Disaster Database.

The effect is an increase in undernourishment of 1 percentage point in the whole of Indonesia, but 4 percentage points within Sumatra (see second part of Table 3.9). This is a relatively frequent event that has a small impact on the nation-wide prevalence of undernourishment, but a much larger effect at the local level. The local prevalence of undernourishment can be a more relevant indicator of the importance of these types of events for policy makers and Indonesian society.

**Table 3.9. Scenario 5: Earthquake in Sumatra**

| | | |
|---|---|---|
| Availability | Production imports and stocks | Rice production in Sumatra declines by 9.8% and rice production in Indonesia declines by 1.7%. |
| Access | Food prices | Increase in consumer price of rice, soybeans and meat by 3.5%, 3.3% and 3.3%, respectively. |
| | Household income and assets | Nominal income of urban and rural household decline by 3.0% and 1.9%, respectively. |

**B. Impacts of Scenario 5: Earthquake in Sumatra**

| | Rate of undernourishment (percentage) | Median calorie intake (kcal per day per capita) | Depth of food deficit (kcal) |
|---|---|---|---|
| Before the shock | | | |
| All Indonesia | 13 | 1 941 | 19 |
| Urban households | 14 | 1 883 | 21 |
| Rural households | 12 | 1 974 | 18 |
| Sumatra | 10 | 1 998 | 15 |
| Urban households | 13 | 1 912 | 20 |
| Rural households | 9 | 2 047 | 13 |
| After the earthquake | | | |
| All Indonesia | 14 | 1 899 | 22 |
| Urban households | 16 | 1 843 | 25 |
| Rural households | 13 | 1 934 | 21 |
| Sumatra | 14 | 1 911 | 21 |

### *Summary of Indonesian food insecurity scenarios*

The reference scenario reflects the current levels of undernourishment with no particular extreme shock in place. Similar circumstances are likely to occur at least every other year and are then estimated to represent the most likely situation (Table 4.8). According to estimates of experts and the discussion in the consultation seminar, the most frequent among the risk scenarios are likely to be crop failures (once every 15 years), followed by earthquakes and high fuel prices (both occurring once every 20 years). Economic crises and rice price spikes are the most unlikely of the six scenarios.

The frequency of local events such as the earthquake in Sumatra requires some explanation. The likelihood is small in a specific location, but the probability of small events in different locations can be larger. The frequency of once in 20 years has been retained following the advice of experts. Although this study focuses on food insecurity risk at the national level, there may be region-specific disasters which require a regional level response.

The analysis of the prevalence of undernourishment in different scenarios needs to account for the estimated frequency of each scenario. The simplest way of doing this is by multiplying the expected prevalence of undernourishment in each scenario by the likelihood of this scenario. This "expected damage" indicator is calculated in column C of Table 3.10. The last column gives information about the regional distribution of undernourishment. In the reference scenario, Maluku is the region with the highest prevalence of undernourishment (24%), nearly double the average for the whole country. For each scenario, the region where the rate of undernourishment rises the most is shown. The numbers given are the regional average *after* the shock, and the size of the increase.

Both the frequency and the prevalence of undernourishment are highest in the crop failure scenario, which makes this scenario most relevant for food security as represented by the "expected damage" indicator. All other scenarios have a smaller prevalence of undernourishment in the range 21-23%. The lowest impact on undernourishment is observed in the scenario of an earthquake in Sumatra, with the prevalence increasing to 14% in the whole country, but with highest impacts on the island itself.

**Table 3.10. Frequency of scenarios and prevalence of undernourishment**

| Scenario | A. Average number of years per occurrence | B. Prevalence of undernourishment | C. Expected increase in undernourishment $(1/A)*(B-B_0)$ | D. Worst local increase in prevalence of undernourishment |
|---|---|---|---|---|
| 0: Reference without shock | 2 | 13 | 0.00 | Maluku 24 |
| 1-b: Rice price spike in international markets | 30 | 23 | 0.33 | Nusa Tenggara 26 (+14) |
| 2: Macroeconomic crisis | 25 | 22 | 0.36 | Java 25 (+11) |
| 3: Increase in international fuel price | 20 | 21 | 0.40 | Sulawesi 20 (+8) |
| 4-a: Crop failure due to plant infestation | 15 | 25 | 0.80 | Java 29 (+15) |
| 5: Earthquake and tsunami in Sumatra | 20 | 14 | 0.05 | Sumatra 14 (+4) |

## 3.3 Analysis of food insecurity risk in Indonesia: Policy analysis (Step 3)

Following the assessment of the risk scenarios, policy analysis was undertaken and the results were discussed in a dialogue with Indonesian stakeholders. As a background to the policy analysis and discussion, this section begins by reviewing the set of relevant regulations and institutions, and the policy programmes, already existing in Indonesia. This information is given in section 3.4.

***Regulations, institutions and programmes***

Recent framework laws relating to agriculture have emphasised food security and food self-sufficiency (Box 3.3). The latest Food Law No. 18/2012 recognises food sovereignty (*kedaulatan pangan*) and food self-reliance (*kemandirian pangan*) as the essential principles of food security. The term *kemandirian pangan* is sometimes translated as 'independence' or 'self-reliance', and at other times as 'self-sufficiency', which causes confusion. The same terms are also used in the Farmers' Empowerment Law of 2013, but with no reference to food security. The Food Law refers to the possibility of a 'food crisis' but without a definition of its meaning. However, the Food Law specifically mentions supply-side threats to food availability. The threats listed in the Law include climate change, invasive species, natural disasters, social disasters, environmental pollution, degradation of land and water, shifts in land use and economic disincentives. This list of threats provides a direct link between the Food Law and food security stability, and is particularly relevant for many of the scenarios that are analysed in this report.

Since 2000, certain trade restrictions including import tariffs, import quotas, import licensing requirements, product registration and stipulated entry ports, have been introduced in Indonesia for various commodities, including sugar, rice, meat, cereals and horticultural products (OECD, 2012b). Even if some of them have been rolled back after negative domestic impacts, they are an indication of a possible anti-trade interpretation of the self-reliance concept in the Food Law (Box 3.3).

The main agricultural policies in place in Indonesia are analysed and discussed in depth in OECD (2012b). The main policies include price support measures, food consumption subsidies and fertiliser subsidies, all of which are somehow linked with Indonesia's self-sufficiency or self-reliance approach to food security. These policies are quantified in the current policy analysis, together with other alternatives like unconditional cash transfers, food vouchers and catastrophic insurance for agriculture.

***Market price support*** constitutes the largest share of agricultural support in Indonesia. It affects several commodities, in particular poultry, rice and sugar, which were the commodities with the highest percentage of gross farm receipts received in the form of transfers (from consumers and government) in 2010-12 (OECD, 2013c). Rice is the focus of the policy analysis in this report because of its importance as a major food staple in Indonesia, being the first source of both energy and protein (see Figure 3.10). As is well known, market price support can only be sustained in the presence of restrictive trade measures. Other agricultural commodities do not receive so much support in Indonesia because they are subject to few trade barriers (the case of soybeans), or because they are hardly tradable (for example, tubers). The costs of market price support measures take the form of efficiency losses and costs for consumers, which are to a great extent invisible because they are not reflected in budgetary costs. Alternative trade policy approaches would focus on building freer and more reliable access to regional and global trade. However, restrictive trade measures remain in place because of strong political economy pressures and lobbying, which make them difficult to reform.

***Rice price support*** is the result of a combination of various trade and domestic measures, most of them managed by the public logistics agency BULOG: trade restrictions, interventions in the rice market and management of a reserve stock system. These three elements are deeply interlinked. Rice imports and exports are subject to strict quantitative controls. A price band (a guaranteed floor price called *purchasing price* for producers and a ceiling price for consumers) is used to trigger the public procurement of rice or the release of rice from the national reserve managed by BULOG. The Food Security Agency monitors the prices, and movement of supplies in or out of the rice reserves is called for if the producer price is more than +/- 15% away from the government purchasing price for two weeks in a row. The Food Security Agency can make a recommendation to buy or sell rice reserves. The recommendation is then approved by the Minister Coordinator for the Economy in concert with other relevant ministers, and the intervention is carried out by BULOG. The only commodity for which there is a proper reserve stock system is rice.

**Box 3.3. Food security legal and institutional framework in Indonesia**

On 18 October 2012, Indonesia's House of Representatives passed a new Food Law No 18/2012. The law replaced the previous one voted in 1996. The Food Law contains provisions regarding staple foods whereby exports will be permitted only after National Food Reserve requirements and domestic food consumption needs have been met, and imports will be allowed only if the government judges that domestic food production plus National food reserves are not sufficient for domestic needs. Other aspects of the Food Law relating to priority for domestic production and consumption of staples reflect what was already the practice. Since 2004, the Indonesian government has controlled imports and exports of rice according to a system of import and export licenses. No specific commodity is mentioned in the Food Law and no specific self-sufficiency objective is stated, only the general principle of self-reliance (*kemandirian pangan*). However, the strategic plan for 2010-14 established production targets for 39 products. For five food commodities (rice, maize, soybeans, sugar and beef), the targeted levels aim to achieve self-sufficiency in meeting forecast consumption.

In terms of food affordability, the government is charged with stabilising supply and the prices of staples, and providing and distributing food staples to the poor. The Law also fixes the objective of improving nutrition and diversifying consumption, and deals with food safety and labelling. Furthermore, it mandates the creation of a new food-security "government institution" reporting to the president and with the task of carrying out the orders of the government relating to the "production, procurement, storing and/or distribution of staple food". The current Food Security Agency under the Ministry of Agriculture is the predecessor of this future "super agency". It currently monitors food availability and food balances, in particular for the five commodities with self-sufficiency targets, and is responsible for early warning systems[1], developing schemes for rice distribution and food reserves. In addition, it analyses prices, in particular the volatility of producer prices and underlying trends in consumer prices, and it promotes food diversification and food safety.

On 9 July 2013, the Indonesian House of representatives passed the new Farmer's Empowerment Law 19/2013. This law is based on the same principles of food sovereignty (*kedaulatan pangan*) and food self-reliance or independence (*kemandirian pangan*) as the new Food Law, but it makes no reference to food security. This law has a chapter on farmers' empowerment focused on education and training, technical assistance, marketing facilities, small farmers' land tenure and access to science and technology. The chapter on farmers' protection allows the use of import tariffs that adjust so as to maintain favourable prices for farmers. In terms of risk management, the Farmers Empowerment Law has several provisions, including the establishment of an early warning system. It allows central and local government to provide compensation after a crop failure, and requires government to assign state or local publicly-owned enterprises to provide agricultural insurance.

The Food Law requires the Indonesian authorities to maintain reserve stocks of rice. For 2015, the target of 10 million tons has been set for public and private reserve stocks combined. There are four levels of public reserve stocks: the national food reserve, provincial government reserves, district government reserves and community food reserves. The Food Security Agency has developed guidelines for managing provincial, district and village reserves. Most of the public reserve stocks are managed by BULOG and are used for two purposes: price stabilisation and disaster relief. In addition to public stocks, private stocks are also held by farm households, industry and traders. According to AMIS, rice reserves in Indonesia at the end of the marketing year have been steadily increasing in recent years, more than tripling from 2 million tons in 2005/06 to 6.3 million tons in 2013/14. According to a joint estimate from the Food Security Agency, BPS and the survey conducted by Sucofindo in March 2011 about 20-25% of the reserves were held by BULOG, another 20-25% by farm households and the rest by the industry and traders. The Coordinating Ministry for the Economy periodically holds meetings with other relevant ministries and decides on reserve interventions by BULOG.

The East Asia Emergency Rice Reserve (currently ASEAN Plus Three Emergency Rice Reserves APTERR) is a coordination exercise based on the national stocks of member countries, with humanitarian aims. Each country earmarks a given amount of rice and cash reserves to be used in the context of APTERR in the case of disasters. It is not intended to be a substitute for trade. The amount earmarked by Indonesia for APTERR is very modest. Most earmarked reserves are from China, Japan and Korea. This reserve has been used on very rare occasions and its use has been confined to natural disasters.

1. For example, work with the WFP on the *Food Security and Vulnerability Atlas*, 2010.

The result of this combination of rice support measures is a domestic price that has been systematically higher than the world price since 2009. This harms access to food by the poor because most of the poor in Indonesia are net buyers of rice. Additionally, although the domestic price is supposedly stabilised, it is not without negative consequences. During a price hike, such as in 2008, exports are not allowed. This impedes the transmission of high international prices to the domestic market, but it also exacerbates the size of the world market spike itself. When world market prices are lower than the minimum purchasing price, imports are controlled or banned. This isolates Indonesia from the world market and impedes the development of economic opportunities and commercial relationships that can be crucial for the quick delivery of imports in the case of urgent need, such as after a disaster like those described in scenarios 4 (crop failure due to a plant infestation) and 5 (earthquake).

***Fertiliser subsidies*** are a major agricultural policy in Indonesia. The subsidies are provided to fertiliser companies, which are state-owned enterprises (SOE) required to sell their products below a Highest Retail Price to small farmers producing on less than 2 hectares. This indirect support only partially reaches farmers due to inefficiencies of SOEs and rent extraction. Cheaper fertilisers should stimulate production and lower consumer prices, but the savings are unlikely to be transmitted to consumers due to the maintenance of a minimum purchasing price for rice. These subsidies are also supposed to increase the economic returns of farm households, but are not targeted to the poor. [22] ***Raskin***[23] is a domestic food aid programme that delivers rice at subsidised prices, prioritising poor or near-poor households. Raskin is the largest assistance programme to the poor in Indonesia, and is thought of as a consumer subsidy to compensate for the negative impacts of price support on access to food by the poor. It provides relatively small amounts of no more than 10 kg/household/month, which is less than 10% of average household consumption. Raskin rice accounts for about 8% of total annual rice sales in Indonesia. Roughly 50% of the entire Indonesian population buys Raskin rice at least once a year. The SUSENAS database contains detailed information about the households that benefit from Raskin. Since the same agency, BULOG, manages both the market price support measures and Raskin, this creates a strong inter-linkage between these measures. BULOG normally obtains the Raskin rice from domestic public procurement or from imports and, in principle, national reserves cannot be used for Raskin.

An alternative way of providing domestic food aid, through food vouchers, is also analysed in this study. It has several advantages: it allows households to spend the subsidy freely on a diversified set of food items, and it breaks the inter-linkage between rice support and domestic food aid[24].

The most important Indonesian social assistance programme after Raskin is the *Bantuan Langsung Tunai (BLT)*, which is one of the largest targeted cash transfer programmes in the developing world. It was first introduced as a compensation to the poor for reductions in fuel subsidies in 2005 and was used for the same purpose in 2008 and 2013. It provides transfers, which change from one year to the next, and are normally described as ***Unconditional Cash Transfers*** (UCT). According to the World Bank (2012), BLT provides temporary protection for poor households with a very lean administrative apparatus, but it is not appropriate for the long-term poverty reduction goal. For reaching this goal, conditional cash transfers, health insurance and school scholarships are considered to be better instruments. In this study, the BLT programme is examined from the perspective of its contribution to alleviating transitory food insecurity.

The transfers generated by current Indonesian policies in recent years are presented in Table 3.11. The amount of price support to rice varies greatly, depending on the international price of rice: it was negative during the price spike of 2008-09, benefiting consumers rather than producers; and it became positive again from 2010 onwards, taxing consumers with higher prices. Unconditional cash transfers under the BLT programme was conceived as an emergency income support measure and, therefore, it too has a very variable budget that adjusts to changing circumstances. Raskin and fertiliser subsidies have had a steadily growing budget, with very similar average expenditure in the last eight years.

**Table 3.11. Selected policy transfers in Indonesia**

| Billion rupiah | 2005 | 2006 | 2007 | 2008 | 2009 | 2010 | 2011 | 2012 | Average 2005-12 |
|---|---|---|---|---|---|---|---|---|---|
| Rice market price support* | -5 927 | 24 180 | 34 329 | -68 679 | -27 484 | 71 126 | 90 469 | 140 055 | 32 259 |
| RASKIN** | 6 357 | 5 320 | 6 584 | 12 096 | 12 987 | 13 925 | 15 267 | 20 926 | 11 683 |
| Fertiliser subsidies** | 2 527 | 3 166 | 6 261 | 15 182 | 18 533 | 18 412 | 16 345 | 13 959 | 11 798 |
| Unconditional cash transfers BLT*** | 4 487 | 18 619 | 0 | 13 966 | 3 844 | n.a. | n.a. | n.a. | 8 183 |

* Estimated transfers to producers according to the PSE methodology. *Source*: OECD PSE Database 2013.

** Budgetary transfers to consumers. *Source*: OECD PSE Database 2013.

*** Budgetary transfers to households. *Source*: World Bank (2012b).

***Agricultural insurance*** is another instrument that aims to help farms and farm households to manage their risk. Insurance programmes that cover significant parts of the agricultural sector are typically highly subsidised and can have a large budgetary cost. Indonesia does not have a large government agricultural insurance programme. Since 2008 there have only been two pilot insurance projects, in the rice and cattle sectors. A stylised rice insurance programme for systemic catastrophic risks is simulated for rice.

The risk of food insecurity implies a financial risk for the government. Food production shocks, natural disasters and import price hikes can have a negative impact on the balance of payments and foreign currency reserves, and they can worsen the country's ability to import food and to implement agricultural and social policy measures in response to risky scenarios. *Market-based mechanisms* can help to manage this governmental risk, and they have been used by governments in other countries. Weather index insurance focuses on production risk associated with weather conditions and was used by the government of Malawi in 2008-11. Call option contracts in the futures markets to lock-in a maximum price of maize have been used by the government of Mexico. Non-physical "over-the-counter" instruments, such as derivatives, could also be used (FAO et al., 2011, p23). All these market tools for managing government risk can potentially complement other policies and ensure their financial viability across all scenarios. Scenarios in which food insecurity is generated by difficulties with the balance of payments were not selected by the risk assessment consultation process. However, given the recent deterioration in Indonesia's external balance (OECD, 2015), it may be interesting to examine how these market tools could be used to promote more stable access to imports and hence manage this type of government risk.

Finally, there are other agricultural programmes that can contribute to food security, such as *ad hoc* payments following natural disasters. However, the size of these programmes is currently small and hence they are not analysed in this study.

### *The impact of selected policies in the reference scenario*

In recent years, the Indonesian government has applied a restrictive trade policy to rice (OECD, 2012b). Imports are limited by quotas, whereas exports are controlled by a strict license system and are, *de facto*, banned. The state agency BULOG manages the market through a monopoly on trade, a system of public procurement and reserve stocks, and a price band (guaranteed floor price for producers and a ceiling price for consumers). As a consequence of this set of measures, the domestic price of rice in Indonesia has been significantly higher than the world price in recent years. In 2010, the price of rice in the world market was 30% lower than the internal price.

The estimated impact of the rice price support policy in the reference scenario (which is the most likely to occur) depends on assumptions about the degree of price transmission between world market prices and domestic prices and the degree of adjustment in food consumption to price changes. Based on the estimated demand system, Table 3.12 shows the simulated impact of the rice support policy in the reference scenario, when these assumptions are allowed to vary.

Table 3.12 shows that price support always increases the prevalence of undernourishment, but the extent to which it does so depends significantly on the assumptions underlying the simulation. With partial price transmission[25] and full dietary adjustment to the new prices, the rate of undernourishment deteriorates by just 2 percentage points, whereas with full price transmission and full dietary adjustments, the rate of undernourishment could increase by as much as 22 percentage points to 33%, due to the higher domestic prices. When households fully adjust their consumption patterns in response to higher rice prices, rice consumption falls by 5%, assuming median elasticities, and the consumption of other food items such as dairy and vegetables increases. The nutritional consequences of these changes for the intake of different nutrients deserve further research. Moreover, the change in food expenditure also squeezes expenditure on non-food items. In the extreme (but unreal) case of no adjustment in consumption patterns, three quarters of the population would qualify as undernourished. This suggests that if dietary patterns do not adjust fully, and if whatever adjustment occurs is not immediate, short-run impacts could be much greater than +2, or +22, extra percentage points. In the rest of the analysis, partial price transmission (when MPS is used) and full dietary adjustment (when price changes are involved) will be assumed. This means that when the reference scenario (with MPS in use) is used as a benchmark, it is assumed that the rate of undernourishment is 2 percentage points above the 'no-policy' reference scenario.

Thus, price support measures have very negative impacts on food security in the reference situation where no shock is involved. As shown below, these negative impacts of the rice price support policy prevail across several other scenarios as well, particularly in all climate or disaster scenarios that reduce domestic yields and production, and with all macroeconomic shocks that have broadly-based negative effects on economic growth. Restrictions on imports increase undernourishment in any domestic food shortage scenario and strongly damage food security.

Figure 3.14 shows the impact of rice price support graphically, according to the alternative assumptions in Table 3.12. The figure intends to show that the rate of undernourishment (i.e. the share of the population below the threshold) is 11%, 13% and 33%, respectively, when there is no price support, price support but partial price transmission, and price support with full transmission. This figure illustrates how policies cause horizontal shifts in the distribution of calorie intake over the population.

**Table 3.12. Reference scenario with no shock: Estimated impacts of rice price support policies**

| | Rate of undernourishment, % | Median calorie intake, kcal | Depth of food deficit kcal |
|---|---|---|---|
| Without price support | 11 | 1 990 | 16 |
| With price support + partial price transmission | | | |
| Assuming full adjustment of dietary pattern | 13 | 1 941 | 19 |
| Assuming NO adjustment of dietary pattern | 26 | 1 726 | 50 |
| With price support + full price transmission | | | |
| Assuming full adjustment of dietary pattern | 33 | 1 680 | 94 |
| Assuming NO adjustment of dietary pattern | 76 | 1 023 | 441 |

**Figure 3.14. Impacts of rice price support on the distribution of calorie intake**

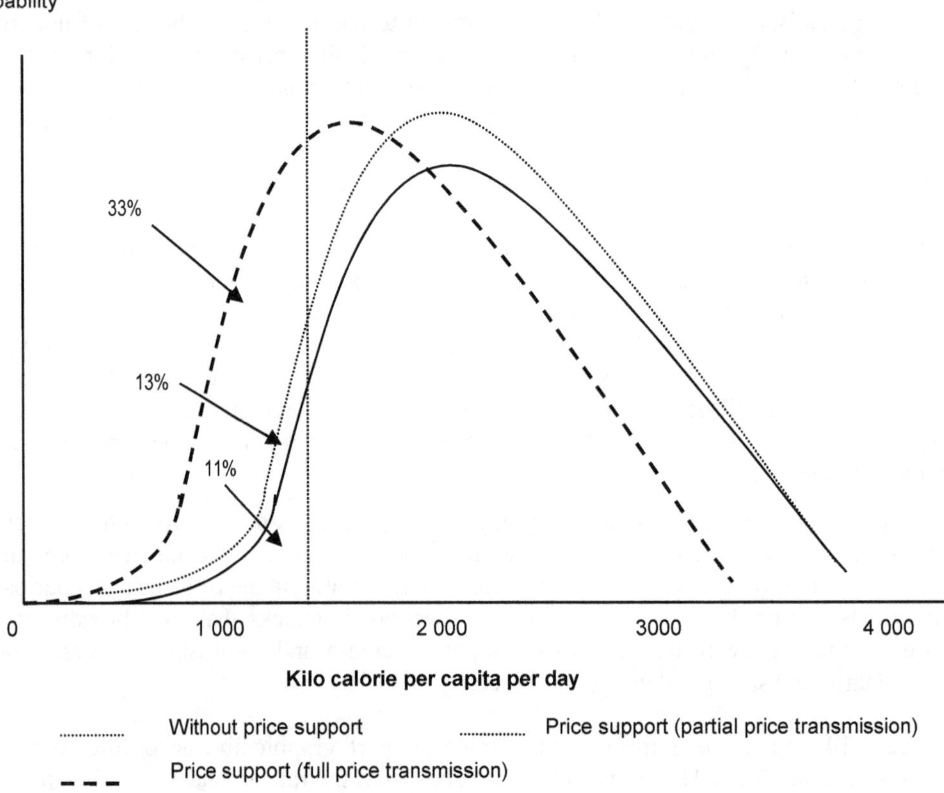


The degree of targeting of both Raskin and BLT is often called into question (World Bank, 2012a, 2012b). The degree of targeting depends on the decisions made at the community level when selecting eligible households (OECD, 2013c). It is often the case that once a community receives support from the government that this support is distributed equally among the households in the community, following an egalitarian social norm. The SUSENAS database includes information on the degree of targeting. Among the poor, 70% of the households receive Raskin (Figure 3.15). But interpreting this figure as good targeting is misleading because more than 40% of the non-poor also receive Raskin and there are recipients of the programme even in the 10% of households with the highest expenditure. Targeting to the undernourished is more difficult because energy intake is less observable than income and consumption. This is why the proportion of the undernourished that receive Raskin rice is 50%, only marginally higher than the proportion of the non-undernourished. The same result applies to the quantity of Raskin rice, which is only significantly higher for poor households, but is not for the undernourished.

In practical terms, the data from SUSENAS show that the relative percentage of population benefiting from BLT across expenditure groups follows a similar pattern, even if Raskin benefits a larger number of households in all income classes. The demand system for all households in the SUSENAS sample is used to simulate the impacts of these social programmes on poverty and undernourishment, assuming the same budgetary outlay for BLT as for Raskin (Figure 3.16).[26] An alternative domestic food aid programme that distributes vouchers to the recipients of Raskin is also simulated. The value of the vouchers per household is equal to the value of the rice subsidy, but the recipient has the freedom to spend it on other staple food items such as grains, tubers, oils and fats. The transaction costs of this voucher programme would normally be smaller than those of Raskin, further increasing its impact on reducing undernourishment.[27]

**Figure 3.15. The recipients of Raskin**

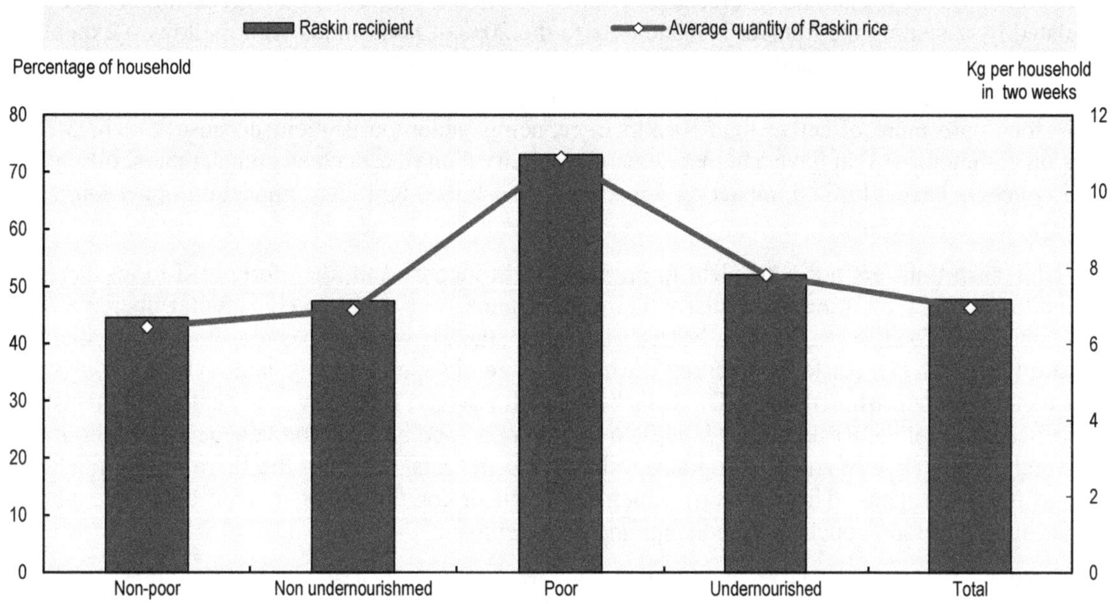


*Source*: SUSENAS database.

**Figure 3.16. The impacts of transfers through Raskin, food vouchers, cash transfers and fertiliser subsidy**

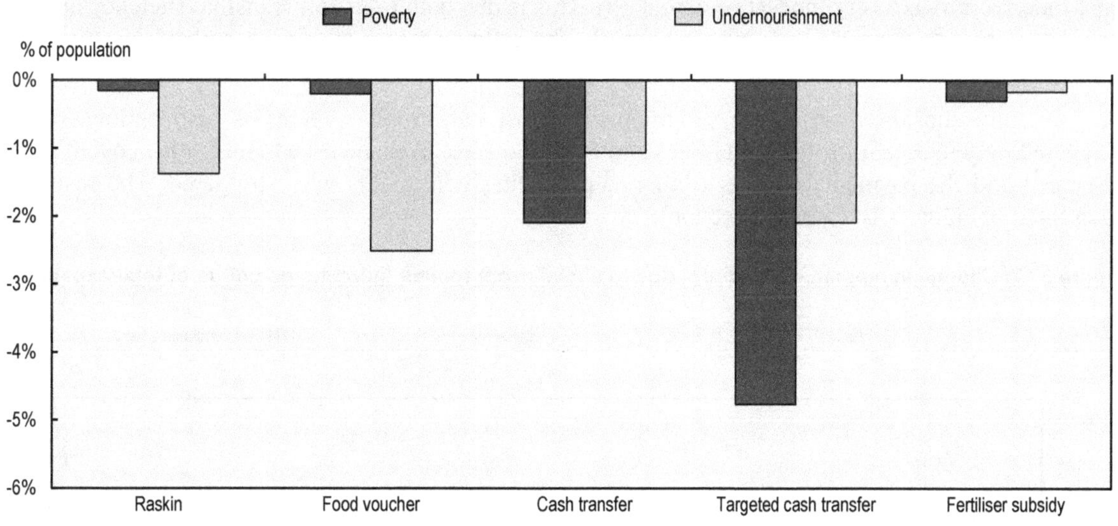


| | Poverty (%) | Undernourishment (%) |
|---|---|---|
| Raskin | -0.2 | -1.4 |
| Food voucher | -0.2 | -2.5 |
| Cash transfer | -2.1 | -1.1 |
| Targeted cash transfer | -4.8 | -2.1 |
| Fertiliser subsidy | -0.3 | -0.2 |

If Raskin were eliminated, the percentage of the poor increases only by a few decimals of a percentage point, while the prevalence of undernourishment increases by more than 1.2 percentage points (Figure 3.16). Raskin is directly subsidising rice consumption and having a particular impact on calorie intake so that it is better targeted to undernourishment than to poverty. The opposite is true for the unconditional cash transfers BLT, which reduce poverty by 2.1 percentage points but undernourishment by less than 1 percentage point. The effect of improving the targeting of BLT is also simulated by assuming the transfer is made only to the 20% of households with the lowest expenditure. This reduces both poverty and undernourishment by significantly more than the current programme, although there are doubts about its feasibility in practice. Finally, food vouchers that can be spent on all staple foods are more effective than Raskin in reducing undernourishment because vouchers can be spent on commodities that have a higher demand elasticity than rice, such as grains, tubers, oils and fats. Food vouchers have a limited impact on poverty (-0.2%) but a significant impact on undernourishment (-2.5%).

BLT cash transfers are equivalent to an increase in income and, therefore, lead to an increase in expenditure across all (and particularly non-food) items, but a relatively smaller increase in the consumption of food items, especially rice and other staples, which have low income elasticities of demand (Figure 3.17). Raskin is focused on making rice cheaper. This explains why Raskin is better targeted to undernourishment and BLT deals better with poverty. However, rice demand is inelastic and a significant part of the Raskin transfer spills over into non-food items. Food vouchers allow households to diversify the source of calories to grains, tubers, oils and fats, reducing the share of the transfer that goes to non-food items. These effects, which are stronger for the lower income classes, explain the greater impact of food vouchers in reducing undernourishment.

Because SUSENAS has no information on fertiliser and insurance subsidies, they have to be simulated in the demand system using a reduced form. The fertiliser subsidy is assumed to have a low transfer efficiency of 15% (OECD, 2001)[28] and it is assumed to be received only by rice-producing households every year. The amount of government expenditure allocated to fertiliser subsidies is similar to that for Raskin (Table 3.11), but the estimated impacts of the fertiliser subsidy policy on both poverty and food security are very modest (Figure 3.16). This is due both to its low transfer efficiency and to its weak targeting to the poor and undernourished. The estimated impact of these subsidies on lowering food prices is small.

A fully subsidised insurance programme against catastrophic events is also simulated. It is triggered only in scenarios 4 (crop failure) and 5 (earthquake). A standard administrative cost of 30% is assumed and the payment is made to all rice-producing households when triggered. The results are presented in Figure 3.18 (see next section).

**Figure 3.17. Change in median expenditure due to social programmes (percentage points of total expenditure)**

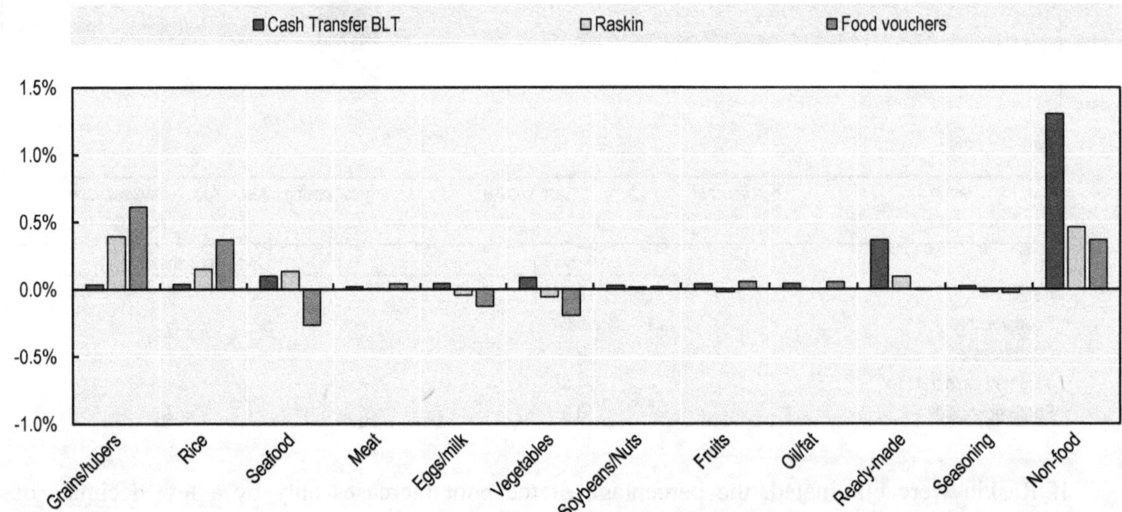

### *The impact of selected policies across the scenarios*

The impacts of rice price support measures on the prevalence of undernourishment in the reference scenario have already been discussed above. Table 3.12 summarises the results conditional on various assumptions, including the conservative case of partial price transmission. The whole set of trade restrictions have a very negative impact on undernourishment not only in the reference scenario but also in all economy-wide risk scenarios such as an economic crisis (2) and an international fuel price hike (3).[29] The impacts are even larger in domestic shock scenarios that generate rice scarcity in Indonesia, such as the crop failure due to pest infestation (4) and the earthquake in Sumatra (5),[30] for which access to imports becomes urgent.

In reality, market price support is combined in a portfolio of policies that also includes unconditional cash transfers BLT, RASKIN and fertiliser subsidies. It could also be combined with crop insurance. The impacts of combinations of programmes (market price support plus one other, in turn) are also displayed in Figure 3.18 for three scenarios. Under the reference scenario, the increase in undernourishment due to price support (1.7 additional percentage points of the population undernourished) is partly mitigated by Raskin or BLT cash transfers, but only marginal improvements are obtained from fertiliser subsidies and there is no improvement from crop insurance. In the price hike scenario, the additional reductions in undernourishment from combining price support with other programmes are relatively small. In the crop failure scenario, crop insurance is the most effective programme for reducing the harm of price support on undernourishment, but even with crop insurance, price support increases undernourishment by 6.9 percentage points. The mitigating effect of crop insurance is not enough to cancel out the negative impact of price support measures on undernourishment.

**Figure 3.18. The impacts of rice price support combined with other policies in different scenarios**

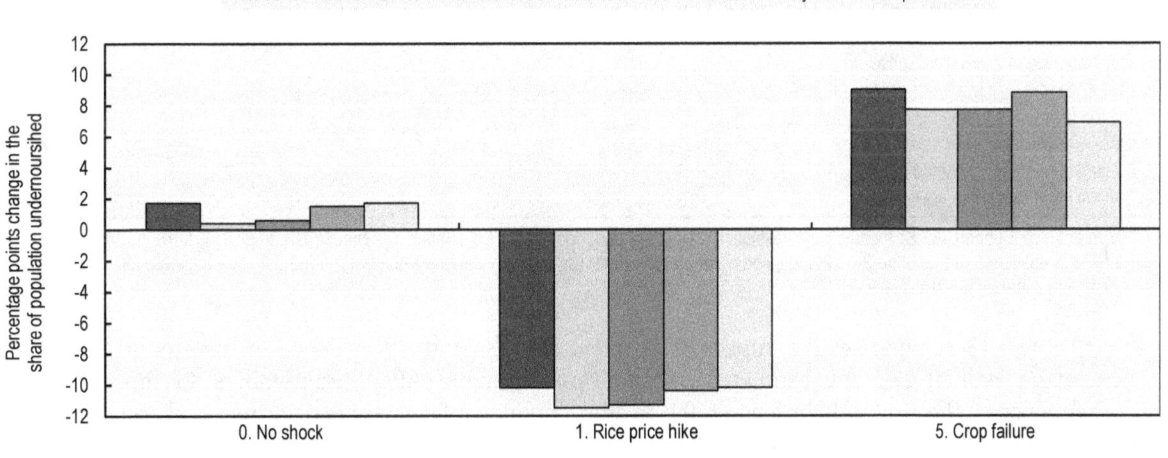


| | 0. No shock | 1. Rice price hike | 5. Crop failure |
|---|---|---|---|
| Alone | 1.7 | -10.2 | 9 |
| With Raskin | 0.4 | -11.5 | 7.7 |
| With cash transfers BLT | 0.6 | -11.3 | 7.7 |
| With fertiliser subsidy | 1.5 | -10.4 | 8.8 |
| With crop insurance | 1.7 | -10.2 | 6.9 |

**Figure 3.19. The impact of selected policies across scenarios**

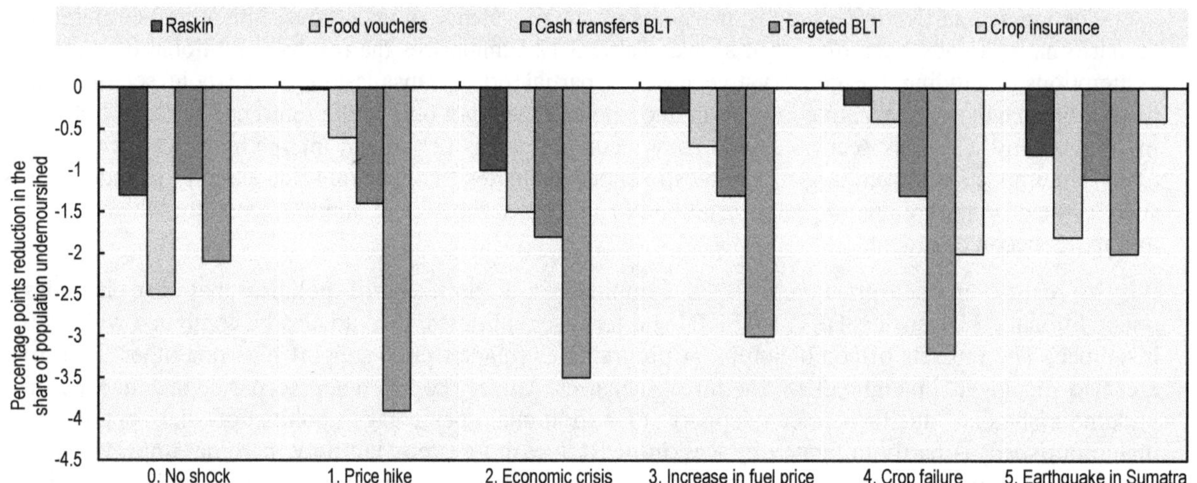


**Table 3.13. Impacts of different programmes on the rate of undernourishment (change in percentage points)**

| Scenario | | Average # of years per occurrence | No policy | Market Price Support (*) | Raskin | Food Vouchers | BLT | Targeted BLT | Fertiliser subsidy | Crop Insurance |
|---|---|---|---|---|---|---|---|---|---|---|
| 0: Reference without shock | | 2 | 0 | 2.0 | -1.3 | -2.5 | -1.1 | -2.3 | -0.2 | 0 |
| 1: Rice price spike in international markets | | 30 | +10.0 | -10.0 | -0.1 | -0.6 | -1.5 | -3.5 | -0.4 | 0 |
| 2: Macroeconomic crisis | | 25 | +9.0 | +2.0 | -1.0 | -1.5 | -2.0 | -3.9 | -0.3 | 0 |
| 3: An increase in international price of fuel | | 20 | +8.0 | +2.0 | -0.3 | -0.7 | -1.4 | -3.3 | -0.2 | 0 |
| 4: Crop failure due to insect or plant disease infestation | | 15 | +3.0 | +9.0 | -0.2 | -0.4 | -1.5 | -3.5 | -0.2 | -2.0 |
| 5. Earth quake and tsunami in Sumatra | Indonesia | 20 | +1.7 | +4.0 | -1.1 | -1.9 | -1.2 | -2.4 | -0.2 | -0.9 |
| | Sumatra | | +3.3 | +6.0 | -0.8 | -1.8 | -1.1 | -2.0 | -0.2 | -0.4 |

* Market Price Support (MPS) means trade restrictive measures plus domestic price support measures. The impact of MPS on the reference scenario is estimated in the range 2 – 22%, depending on assumptions; the most conservative number 2% (partial price transmission + full dietary adjustment) is presented in this table.

Table 3.13 summarises the impact of different shocks on the prevalence of undernourishment (in percentage points) and the estimated capacity of different policy programmes to mitigate these consequences. The first column shows the average number of years per occurrence of the conditions described in the scenario, and is hence inversely related to the frequency. For example, a rice price spike in international markets is expected on average every 30 years, indicating an expected frequency (probability) of 0.033. The other columns of the table show the change in the rate of undernourishment, relative the situation where the scenario occurs but no measure is adopted. The change in the rate is given in percentage points of the population.

Table 3.13 shows that Raskin, cash transfers BLT, food vouchers and the fertiliser subsidy improve the undernourishment situation in all scenarios, but targeted BLT transfers do so the most, while the impacts of fertiliser subsidies are only marginal. Trade restrictions accompanied by market price support worsen the undernourishment situation in all scenarios, except for the least frequent scenario of a rice price spike in international markets. Considering the significant adverse impact of this policy package on undernourishment in all scenarios, including the reference scenario, apart from scenario 1, rice price support cannot be considered a robust policy for addressing transitory food insecurity. Crop insurance is unable to improve the prevalence of undernourishment except in the disaster scenarios 4 (crop failure) and 5 (earthquake).

The impacts of different policy instruments need to be evaluated keeping in mind their consequences under all possible scenarios as well as the likelihood of each scenario. The positive and negative impacts in Table 5.3 should be evaluated in the light of the corresponding assumed probabilities for each scenario. These probabilities are the frequencies obtained as the inverse of the expected number of years per occurrence given in the first column of Table 3.13. These probabilities are used to calculate a weighted average of impacts across all scenarios, which are presented in Table 3.14.

There is a lot of policy relevant information summarised in this table. The Raskin and BLT assistance programmes are able to reduce the shock to undernourishment by 1.0 and 1.2 percentage points on average. This means a considerable reduction of the impacts of the shocks on undernourishment by more than half, on average. Food vouchers, with an average reduction of 2 percentage points, perform better Raskin, and targeted cash transfers with reduction of 2.6% are even more effective on average. In fact, targeted cash transfers are the only programme that is able to fully overcome the shock on food security, reducing the prevalence of undernourishment beyond the increase due to the shocks on average. On the contrary, rice price support, with its associated trade restrictions, increases the shock on undernourishment by 2.4 points on average, more than doubling it from 2 points to 4.4. This is due to its negative effects across most of the scenarios, with the exception of scenario 1. Fertiliser subsidy and crop insurance have only small impacts on reducing the prevalence of undernourishment on average (0.2 percentage points).

The OECD Framework for the Analysis of Transitory Food Insecurity set out in section 2 includes the use of portfolio analysis to identify best policy practices for improving transitory food security. This means analysing the impact of policies from the double perspective of their average impacts on the rate of undernourishment and the variability of these impacts. For this purpose, the average impact of the different policies across scenarios is not enough and some estimation of the variability is needed. The last row in Table 3.14 displays the change in the (probability-weighted) average standard deviation of the impact of the shock when a given policy is in place. Cash transfers, in particular if they are targeted, are the most effective policy for reducing the variability of the impacts of the shocks on the rate of undernourishment. Fertiliser subsidies and crop insurance have a very small variability of impact, since their impacts tend to be small in all or most cases. Raskin and food vouchers do not reduce the variability due to the shocks because these policies are more effective in the reference scenario in which the recipients are well identified as compared to scenarios under an external shock. Rice price support also increases the variability in the impacts of the shocks. Overall, rice price support measures are increasing the risk of undernourishment.

**Table 3.14. Average effect of policies on the shocks to the rate of undernourishment and its standard deviation**

| | Impact without policy | Changes in rate of undernourishment due to policy, percentage points | | | | | | |
|---|---|---|---|---|---|---|---|---|
| | | Market Price Support | Raskin | Food Vouchers | Cash Transfers - BLT | Targeted BLT | Fertiliser subsidy | Crop Insurance |
| Average impact | 2.0 | +2.4 | -1.0 | -2.0 | -1.2 | -2.6 | -0.2 | -0.2 |
| Standard deviation of impact | 3.3 | +0.53 | +0.34 | +0.65 | -0.20 | -0.47 | -0.03 | -0.01 |

## *Notes*

1. Another important source of information for local analysis of transitory food insecurity in Indonesia is the *Food Security and Vulnerability Atlas of the World Food Program* (WFP 2009, 2012). It provides detailed information on the vulnerability to food security risk in different districts. The *Cost of Diet Methodology* (Baldi et al., 2013) allows to estimate the local monetary needs for a good diet.

2. This report defines nine regions for seven groups of islands, separating Jakarta and Bali from Jawa Island.

3. All data and figures in this section on poverty and undernourishment and in the next section on nutritional outcomes are based on data from SUSENAS database and calculations by the OECD Secretariat.

4. BPS sets the regional poverty line and food poverty line for urban and rural households separately. When the household is located in an urban area, its total expenditure is compared with the urban (food) poverty line to judge whether it is (food) poor.

5. A household is classified as farm household when the occupation of the household head is an employer in agriculture (including self-employment).

6. Figure 3.6 shows an increase in the rate of undernourishment from 2008 to 2009. The FAO record a reduction in the same period. This is the result of the different threshold and distribution between our calculations and FAOs. Additionally the FAO publishes its results in the form of three year moving averages, which captures better trends than variability.

7. Child malnutrition is one of the more extensively used indicators of nutritional outcomes. According to FAO the percentage of children under five years of age who are underweight has been reduced in Indonesia from 30% in 1992 to 25% in 2000 and 19% in 2010. This information is not available in SUSENAS database.

8. The large shares of rice in calorie and protein intake are confirmed with similar shares of supply calculated by FAO.

9. See Warr and Yusuf (2014). The quantification exercise consisted in estimating the consequences of each scenario in terms of income for each rural and urban centile household and in terms of relative prices.

10. Although there are several indicators of food insecurity, some of which have been discussed and presented in Chapter 1, the analysis in Chapters 2 and 3 is focused on the prevalence of undernourishment. The demand simulation model allows similar analysis and quantifications for other indicators, such as poverty, food poverty and those indicators based on other nutrients (e.g. protein intake).

11. This section draws heavily on a consultant paper by Peter Warr from the Australian National University.

12. By contrast, in the Philippines, the peak monthly retail price of rice in June 2008 was 62% higher than the annual average retail price in 2007 following the international price hike of about 200% that occurred between September 2007 and May 2008.

13. The IFLS is a longitudinal survey in Indonesia, whose sample is representative of about 83% of the population. The first wave of the survey was conducted in 1993-4, and the fifth wave (IFLS-5) will be in the field in 2014-15. It is conducted by RAND.

14. A third version of this scenario was also run, but not retained for discussion, in which an *ad valorem* tariff of 35% was used (which restricts the initial volume of imports of rice by the same amount as the quantitative restriction, but allows some price transmission to the domestic market). In this version, the rate of undernourishment increases, but by less than in scenario 1-b.

15. These price transmission parameters were derived from simulations using the INDONESIA-E3 general equilibrium model. The almost zero price transmission in scenario 1-a is due to policy, that is import and export bans. The almost 50% price transmission in scenario 1-b reflects both policy aspects (import restrictions) and other physical or economic factors influencing price transmission.

16. The extent of the food deficit indicates how many calories per capita would be needed to lift the undernourished from their status, everything else being constant. The numbers do not match with the FAO numbers because we use real household data while FAO calculations are based on assumed distributions of calorie intake.

17. Warr and Yusuf (2014) also examined the impact of the import quota on food insecurity using a general equilibrium modelling framework. They found that the reduction in permanent poverty incidence that would result from lower domestic rice prices if the restriction was eliminated (assuming the export restrictions remain operative) far outweighs the increase in poverty that was averted by preventing the transmission during the international price surge. Similar results could be expected with regard to food security.

18. BPH is a small, fast-breeding insect that invades the stalks of rice plants. Its feeding causes "hopper burn", but it also carries two damaging viral diseases of rice (ragged stunt virus and grassy stunt virus) which can be as damaging to the rice crop as the feeding of the BPH itself.

19. See http://ricehoppers.net/2014/01/the-threat-to-rice-production-in-java-in-2014-by-planthopper-pest-outbreaks/.

20. The BPH is also a major pest elsewhere in the region, including Thailand, Cambodia and Vietnam. In the mid-1980s, outbreaks of BPH infestation occurred throughout Southeast Asia. Repetition of this kind of event could cause an increase in world rice prices at the same time as the fall in production in Indonesia. However, the likelihood of this coincidence of events is reduced by the fact that the BPH is not mobile in the short-run between Indonesia and mainland Southeast Asia. Furthermore, unlike Indonesia, other Southeast Asian countries, such as Viet Nam, have restricted overuse of pesticides, reducing the probability of an outbreak in their country.

21. TERM (The Enormous Regional Model) is a "bottom-up" CGE model of Australia, which treats each region as a separate economy. A version of TERM exists for Indonesia. See Wittwer, 2012, and http://www.copsmodels.com/term.htm

22. OECD (2014) provides more details of this policy and quantitative estimates of the impacts of the fertiliser subsidy on agricultural production and income in Indonesia.

23. Raskin means "rice for the poor" in Indonesian. Formally, the programme is called "Beras Bersubsidi bagi Masyarakat Berpenghasilan Rendah".

24. BULOG manages rice imports and the purchase of rice from domestic producers. If the subsidised rice program Raskin were converted to a voucher program, it would reduce the necessity for BULOG to import rice or to purchase it internally. Various Indonesian experts consider that breaking the interaction between these policy measures would be an important reform step for Indonesia.

25. The partial price transmission scenario is simulated according to the results of the general equilibrium model for Indonesia known as INDONESIA-E3 (Warr and Yusuf, 2013).

26. Per household cash transfer was calculated so that total budget including 5.4% programme administrative cost equals 2010 budget outlay of Raskin programme.

27. In the simulations in this paper the transfer efficiency of Raskin and food vouchers is assumed to be the same. The estimated gains on reducing undernourishment are just due to the possibility of using the subsidy for food items with higher demand elastic than rice.

28. The transfer efficiency of a policy measure is the percentage of the government disbursement that reaches the intended recipients. Transfer efficiency is typically less than 100% because of "leakages" to unintended recipients (e.g. economic rent captured by 'middlemen'), and deadweight losses (=economic inefficiencies).

29. The same impact on undernourishment of +2 percentage points is applied to scenarios 2 and 3 in the analysis in this section.

30. The impact of the earthquake in Sumatra will be smaller than the broader crop failure that affects all Indonesia. The increase in the prevalence of undernourishment is assumed to be +4 percentage points in Indonesia and +6 in Sumatra, compared to the estimated +9 in scenario 4 (Figure 3.18).

## Annex 3.A

### Estimation of a household food demand system in Indonesia[1]

The Almost Ideal Demand System (AIDS) is a standard economic model for representing a demand system, first proposed by Deaton and Muellbauer (1980). The strengths of this model are that 1) it is based on utility maximisation, 2) it gives a first-order approximation to any demand system, and 3) by means of restrictions on parameters, it embodies and can test ideal properties of demand such as homogeneity of degree zero in prices and symmetry of reactions to price changes across commodities. A typical equation of the AID system is written as

$$w_{ih} = \alpha_i + \sum_j \gamma_{ij} lnp_j + \beta_i ln\{^{x_h}/_P\}, \quad i = 1,2,\ldots,I \qquad [1]$$

where $w_{ih}$ is the share of expenditure for commodity $i$ for household $h$, $p_j$ is the price of commodity $j$, $x_h$ is total expenditure for household $h$ , and the price index, $P$, is given as

$$lnP = \alpha_0 + \sum_k \alpha_k lnp_k + \frac{1}{2}\sum_k \sum_j \gamma_{kj} lnp_k lnp_j. \qquad [2]$$

This specification is interpreted as follows: the share of expenditure on commodity $i$ is a function of the prices of all commodities and *real* expenditure. The parameters $\alpha_i, \gamma_{ij}$, and $\beta_i$ are, respectively, the share-specific constant, price effect and expenditure (or income)[2] effect. The more familiar properties of demand functions (price and income elasticities) are functions of these share parameters and other data. For this system to be consistent with utility maximisation, three sets of restrictions on parameters must hold, namely:

$$\sum_{i=1}^{n} \alpha_i = 1, \sum_{i=1}^{n} \gamma_{ij} = 0, \sum_{i=1}^{n} \beta_i = 0,$$

$$\sum_{j=1}^{n} \gamma_{ij} = 0,$$

$$\gamma_{ij} = \gamma_{ji}.$$

The first set of restrictions is the "adding-up" property, which ensures that the expenditure shares sum to one $(\sum w_i = 1)$. The second set of restrictions ensures that each demand function is homogeneous of degree zero in prices and expenditure (income). The third set of restrictions implies Slutsky symmetry. Together these properties reflect the underlying assumption that the consumer is a rational decision maker. These restrictions can either be imposed during estimation, or tested post-estimation.

This system of equations must be estimated together since the disturbance terms for each share are inevitably correlated with each other due to the adding-up property, and various parameters (the αs and γs) appear in all equations.[3] To avoid nonlinearity of the system, Deaton and Muellbauer (1980) replaced the price index, defined in [2], by Stone's price index, namely:

$$ln\, P^* = \sum_{i=1}^{n} w_i\, ln\, p_i. \qquad [3]$$

This specification is called Linear Approximate AIDS (LA-AIDS) and allows linear estimation by a system estimator such as the Seemingly Unrelated Regression (SUR) estimator.

### *Estimation by sub-samples*

Household consumption will react differently to price and income changes according to the household's food preferences, its access to commodities, its income level, and its own countervailing measures such as own-production. Therefore, the total sample of households is subdivided by region, by expenditure class, and by farm and non-farm type, in an attempt to capture systematic parameter differences between these sub-groups.

Each region in Indonesia differs in terms of dietary culture and economic aspects such as industry structure and market infrastructure. For this study, Indonesia is divided into four areas according to similarity of consumption pattern and geographical closeness: Sumatra, Jawa (including Jakarta and Bali), Sulawesi/Kalimantan, and Papua/Malku/Nusa Tenggara.

To allow for different price responses for farm and non-farm households, the total sample is also divided between these two types. 30% of households in SUSENAS are identified as farm households, although the proportion varies by region. Part of the food that farm households consume is produced on their own farm. Therefore, both purchased and own-produced rice are summed in the estimation for farm households, while in the case of non-farm households all rice is purchased. An increase in the price of a commodity produced on the farm creates an incentive to expand market sales and reduce own-consumption. Thus, the estimated demand parameters are expected to differ for farm households due to capture their simultaneous production and consumption decisions. For non-farm households with a positive value for own-consumption of food commodities, the calorie intake from own-production is treated as constant.

The data show that, for some commodities such as meats, there is a non-linear relationship between the share of expenditure on that good and total expenditure. That is, the *ceteris paribus* increase in the expenditure share becomes smaller as one moves to higher income levels. Banks et al. (1997) proposed a method for estimating nonlinear curvature of the Engel curve by adding quadratic term in real income to each equation [1] of the system. Instead of applying this method, which increases the calculation burden, the sample is sub-divided by expenditure classes, allowing the average differences in expenditure parameters with expenditure level to be entirely data-driven. The sub-division by expenditure class is done differently for farm and non-farm households. Non-farm households are divided into five sub-samples by expenditure quintile. Farm households are divided into two sub-samples: those with expenditure less than the 40$^{th}$ expenditure percentile and those with expenditure between the 40$^{th}$ and 100$^{th}$ percentile.[4]

### *Dealing with zero expenditure observations*

The issue of zero expenditure shares for some commodities has received substantial attention in the literature. This issue is crucial when using microdata, which tend to contain numerous zero shares because of corner solutions to consumption decisions, inaccuracy of data processing and missing responses by households. Our SUSENAS data have a non-negligible share of zeros: while only 3% of the households report no rice consumption, 40% do not consume grains & tubers and 60% do not consume meat. The existence of zero shares creates bias in the estimated parameters.

Many empirical applications of the AIDS model confronting this issue follow Shonkwiler and Yen (1999), who offered a constant two-step (CTP) estimation procedure to correct for sample selection estimation biases. They consider the following system of equations, explicitly separating the decision on *whether* a household consumes a particular good from that of *how much* of it is consumed. The model is re-specified as

$$w_{ih} = d_{ih}\, w_{ih}^{*} \qquad [4]$$

The participation equation (first-step equation) is

$$d_{ih}^{*} = z_{ih}'\delta_i + v_{ih}, \qquad [5]$$

where

$$d_{ih} = \begin{cases} 1 & if\ d^*_{ih} > 0 \\ 0 & if\ d^*_{ih} \leq 0 \end{cases}$$

The amount equation (second-step equation) is

$$w^*_{ih} = y'_{ih}\beta_i + \varepsilon_{ih}. \qquad [6]$$

The subscripts, $i$ and $h$, denote the $i$th equation of the demand system of household $h$. $w$ is expenditure share for a particular good and $d$ is a binary variable indicating whether or not consumption occurs. The asterisk indicates a latent variable, which is not observed in the data. $y$ and $z$ are exogenous regressors. Equation [6] corresponds to the LA-AIDS equation.

Following Type I Tobit estimation (Wooldridge, 2002), equation [5] is estimated as a Probit model, to obtain the inverse Mills' ratio $\frac{\phi(z'_{ih}\delta_i)}{\Phi(z'_{ih}\delta_i)}$. Inserting this value into the second-step equation corrects for the bias caused by the presence of zero expenditure. The second-step equation is therefore defined as

$$w_{ih} = \Phi(z'_{ih}\delta_i)x'_{ih}\beta_i + \sigma_i\phi(z'_{ih}\delta_i) + \xi_{ih} \qquad [7]$$

where $\xi_{ih} = w_{ih} - E(w_{ih}|\ x_{ih}, z_{ih})^5$. $\Phi$ and $\phi$ are the cumulative distribution function (CDF) and the probability density function (PDF) of the normal distribution. $\sigma$ is a parameter to be estianted and $\xi$ is the disturbance term.

Shonkwiler and Yen (1999) and Drichoutis et al. (2008) argue that the variance of the error term $\xi_{ih}$ is inevitably heteroskedastic by construction, which renders the estimator from the Shonkwiler and Yen procedure inefficient (although it does not imply bias for the parameter estimates). One solution to this problem would be to use variance-weighted regression. However, estimation of variance is not straightforward. Thus, most relevant work (e.g. Tafere et al., 2010) uses robust standard errors, which is the method followed here.

Since the model in [7] is scaled by the CDF, the expenditure shares do not sum to 1. The model is estimated by dropping one equation (the share of 'other foods'), implying that the share of expenditure on other foods is 1 minus the sum of other expenditure shares (Yen et al., 2003).[6] The relationships between elasticities and expenditure shares arising from the budget constraint are used to calculate the price and expenditure elasticities of the demand for other foods.[7]

***Fixed-effects estimation to control for heterogeneity in households***

Unlike other studies such as Tafere et al. (2010) and Flordeliza et al. (2013), the analysis reported here uses panel-data estimation to make the best use of the three-year panel survey to control for unobserved household-specific effects. Specifically, the within estimator (or fixed-effects estimator) is used to estimate the CDF-augmented LA-AIDS model given in equation [7]. Household-specific effects and time-specific effects are incorporated in equation [7], denoted by the use of the time subscript $t$ to obtain the estimating equation given by [8]:

$$w_{iht} = \Phi(z'_{iht}\delta_i)x'_{iht}\beta_i + \sigma_i\phi(z'_{iht}\delta_i) + \alpha_h\Phi(z'_{iht}\delta_i) + \Phi(z'_{iht}\delta_i)d'_t + \xi_{iht}. \qquad [8]$$

The within estimator is applied to equation [8], which implies estimating

[9]

$$w_{iht} - \overline{w_{iht}} = (\alpha_i + \alpha_h)\left[\Phi(z'_{iht}\delta_i) - \overline{\Phi(z'_{iht}\delta_i)}\right] + \sum_j \beta_{ij}\left[\Phi(z'_{iht}\delta_i)lnp_{jht} - \overline{\Phi(z'_{iht}\delta_i)lnp_{jht}}\right]$$

$$+\gamma_i\left[\Phi(z'_{iht}\delta_i)RE_{ht} - \overline{\Phi(z'_{iht}\delta_i)RE_{ht}}\right] + \sigma_i\left[\phi(z'_{iht}\delta_i) - \overline{\phi(z'_{iht}\delta_i)}\right]$$

$$+\rho09_i\left[\Phi(z'_{iht}\delta_i)d09 - \overline{\Phi(z'_{iht}\delta_i)d09}\right]+\rho10_i\left[\Phi(z'_{iht}\delta_i)d10 - \overline{\Phi(z'_{iht}\delta_i)d10}\right] + \zeta_{iht},$$

where variables with a bar indicate the three-year average, $lnp_{iht}$ is log of price of good $i$ for household $h$ at time $t$, $RE_{ht}$ is real expenditure for household $h$ at time $t$, $d09$ and $d10$ are dummies for 2009 and 2010, and $\zeta_{iht} = w_{iht} - \overline{w_{iht}} - \left[E(w_{iht}|\, x_{iht}, z_{iht}) - \overline{E(w_{iht}|\, x_{iht}, z_{iht})}\right]$. Equation [9] is estimated by the SUR estimator with the equation for the composite commodity 'other foods' dropped.

### *Data construction*

Information on expenditure shares and total expenditure comes from the SUSENAS panel. In this study, thirteen commodities/commodity groups are distinguished: rice, grains & tubers, seafood, meats, eggs & milk, vegetables, soybeans & nuts, fruits, oils & fats, delicatessen items & drinks, seasoning (mostly chili pepper, salt, sugar), other foods, and one non-food aggregate.

For composite commodity groups (that is, all groups except rice), price index data from the Consumer Price Indices from 2008 to 2010 collected by BPS are used rather than weighted averages of unit values.[8] The data source provides price indices (2007=100) for the eleven food aggregates plus non-foods in 66 cities covering all 33 provinces in Indonesia. The price data from each city are assumed to apply to all households located by SUSENAS within that same administrative zone. In the case of rice, price is calculated by dividing the value of rice purchased by quantity of rice purchased in kilograms, and averaging the unit value at the level of the administrative zone.

### *Estimation*

The system of demand equations defined in [1] is estimated using the two-step procedure given in equations [5] and [6]. Estimation is carried out for each sub-sample for the period 2008-10. Around 50% of the estimated parameters are significant at the 10% level. In this study, the significance of coefficients may have been affected by insufficient variation in regional prices. Values of R-squared were low, ranging between 0.001 and 0.7; however, it must be borne in mind that this is to be expected when fitting models with relatively few parameters to a large number of observations. In such cases, low R-squared values can nevertheless be associated with highly significant F-values for goodness-of-fit.

Most of the coefficients for own-price effects on share, $\gamma_{ii}$, are significantly positive except for meats. This implies that with real income fixed, a *ceteris paribus* increase in own-price expands the expenditure share for most goods.[9] Coefficients for the effect of a real expenditure change on expenditure share, (the $\beta_i$), are mostly positive for grains & tubers, meats, eggs & milk, fruits, prepared food (delicatessen), and non-food goods. This is consistent with the idea that these are luxury goods, whose demand increases more than proportionately with an increase in real income, while the rest are staples or ordinary foods. Furthermore, non-food has largest income response parameter among the luxury goods, indicating that income growth will increase the relative expenditure share of the non-food category.

### *Calculation of elasticities*

Elasticities of demand measure the percentage change in demand caused by a 1% change in a particular price or in income, holding everything else constant. They are very important information for consumer analysis and policy design since they make it possible to simulate consumers' reactions to external shocks (e.g. like income and price changes due to natural disasters) and policies. A virtue of the AIDS model is that it implicitly provides elasticities of demand for each household, which are derived from the estimated model parameters and information on total expenditure and expenditure shares.

The formula for the *uncompensated (Marshallian)* price elasticities is as follows[10],

$$\varepsilon_{i,j} = -\delta_{ij} + \frac{\gamma_{ij}}{w_i} - \beta_i \frac{w_j}{w_i} \qquad [10]$$

where $\delta_{ij} = 1\ if\ i = j$ and $\delta_{ij} = 0$ otherwise. $\varepsilon_{i,j}$ is the own-price elasticity of demand. The signs of the own-price elasticities are expected to be negative, that is, an increase in the rice price reduces the demand for rice (except in the textbook case of a Giffen good). $\varepsilon_{i,j}$ $(i \neq j)$ is a cross-price elasticity of demand. Cross-price elasticities are negative when the two commodities in question are complementary and they are positive when the two goods are substitutes.

The expenditure elasticities of demand are given by

$$\varepsilon_i = \frac{\beta_i}{w_i} + 1\,. \qquad [11]$$

Expenditure elasticities can be interpreted as income elasticities under the assumption that savings are a constant proportion of income. A summary of the estimated elasticities for each of five expenditure groups for Indonesia in 2010 is presented in Tables 3A.1 and 3A.2.

## *Notes*

1. This econometric analysis was undertaken in collaboration with Hiroaki Ogawa, Osaka University.
2. Expenditure effects are often interpreted as income effects, implying either that all income is spent on commodities or that expenditure on commodities (and hence also saving) is a constant share of income.
3. Deaton and Muellbauer (1980, p316 and Appendix) suggest exogenously setting $\alpha_0$ equal to minimum expenditure in each sub-sample. The parameter is then interpreted as the expenditure required to achieve a minimum necessary standard of living when prices are equal to unity.
4. The reason for using only two expenditure classes for farm households is the smaller number of households of this type. If more classes were used, some regional sub-samples would become too small for econometric estimation.
5. The conditional mean of $w_{ih}$ is given as the probability-weighted sum of the mean conditional on consumption; $E(w_{ih} | x_{ih}, z_{ih}) = P(d_{ih} = 1 | z_{ih}) E(w_{ih} | x_{ih}, z_{ih};\ \upsilon_{ih} > -z'_{ih}\delta_i) = \Phi(z'_{ih}\delta_i) x'_{ih}\beta_i + \sigma_i \phi(z'_{ih}\delta_i)$. See Wooldridge (2002, p672).
6. Drichoutis et al. (2008) argue that this method is *ad hoc* but practical. They also mention that the whole system can be estimated as it is without dropping one equation (there is no singularity when the adding-up property is not imposed), but that in this case the shares do not sum to 1, thereby causing model inconsistency.
7. For details, see Yen et al. (2003, p460, footnote 9).

8. Using unit values for consumption analysis has certain economic and statistical problems. See Deaton (1988).

9. This implies that only meats have a *compensated (Hicksian)* demand price elasticity smaller than one. The coefficients on own-prices in the AIDS model $\gamma_{ii}$, are negative when *compensated* own-price elasticity, $\eta_{i,p_i} = -\frac{p_i}{q_i(p)}\frac{\partial q_i(p)}{\partial p_i}$, is larger than unity and positive otherwise since $\frac{\partial w_i}{\partial ln p_i} = \gamma_{ii} = w_i(1 - \eta_{i,p_i})$. The compensated elasticity indicates pure substitution between goods caused by change in their relative price with real income unchanged.

10. This formula gives an approximation of the exact elasticities. Green and Alston (1990) found that this formula approximates elasticities calculated with true formula very well and there is little difference. For simplicity of calculation of elasticities, we use this formula.

## *Estimated elasticities*

**Table 3A.1. Own-price elasticities of demand for Indonesia in 2010**

| 0-20 | | | | | | | 60-80 | | | | | |
|---|---|---|---|---|---|---|---|---|---|---|---|---|
| Variable | Obs | Mean | Std. Dev. | Min | Max | | Variable | Obs | Mean | Std. Dev. | Min | Max |
| e1p1 | 2 643 | 0.413 | 2.519 | -4.994 | 4.996 | Grains | e1p1 | 4 085 | -2.698 | 1.190 | -4.997 | 4.884 |
| e2p2 | 11 084 | -0.389 | 0.316 | -0.827 | 4.849 | Rice | e2p2 | 11 652 | 0.063 | 0.660 | -0.908 | 4.883 |
| e3p3 | 8 718 | -0.364 | 0.983 | -4.758 | 4.922 | Seafood | e3p3 | 10 951 | -0.712 | 0.673 | -4.890 | 4.985 |
| e4p4 | 1 511 | -1.865 | 1.409 | -4.998 | 4.980 | Meat | e4p4 | 5 442 | -1.775 | 1.107 | -4.975 | 4.451 |
| e5p5 | 6 308 | -1.601 | 1.258 | -4.994 | 4.973 | Eggs/Milk | e5p5 | 10 724 | -1.135 | 0.934 | -4.877 | 4.991 |
| e6p6 | 11 360 | -0.760 | 0.464 | -4.382 | 4.899 | Vegetables | e6p6 | 11 896 | -1.118 | 0.258 | -4.566 | 3.619 |
| e7p7 | 8 383 | -1.062 | 0.656 | -4.987 | 4.908 | Soybeans/Nuts | e7p7 | 9 755 | -0.944 | 0.357 | -4.612 | 4.870 |
| e8p8 | 5 175 | -1.433 | 1.016 | -4.982 | 4.560 | Fruits | e8p8 | 9 408 | -0.720 | 0.647 | -4.762 | 4.803 |
| e9p9 | 10 903 | -0.851 | 0.747 | -4.678 | 4.982 | Oil/Fat | e9p9 | 11 796 | -0.897 | 0.357 | -4.930 | 4.030 |
| e10p10 | 11 196 | -1.267 | 1.051 | -5.000 | 4.903 | Ready-made | e10p10 | 12 301 | -1.172 | 0.666 | -4.999 | 4.497 |
| e11p11 | 11 511 | -1.056 | 0.892 | -4.989 | 4.994 | Seasoning | e11p11 | 11 881 | -0.731 | 0.816 | -4.966 | 4.909 |
| e12p12 | 6 726 | -1.155 | 0.648 | -4.982 | 2.293 | Non food | e12p12 | 10 171 | -0.688 | 0.715 | -4.979 | 4.949 |
| e13p13 | 12 150 | -0.956 | 0.860 | -4.984 | 4.980 | Other food | e13p13 | 12 433 | -1.206 | 0.549 | -4.916 | 1.643 |

| 20-40 | | | | | | | 80-100 | | | | | |
|---|---|---|---|---|---|---|---|---|---|---|---|---|
| Variable | Obs | Mean | Std. Dev. | Min | Max | | Variable | Obs | Mean | Std. Dev. | Min | Max |
| e1p1 | 2 719 | -2.367 | 2.038 | -5.000 | 4.983 | Grains | e1p1 | 3 845 | -2.887 | 1.623 | -4.999 | 4.993 |
| e2p2 | 11 613 | -0.331 | 0.375 | -0.822 | 4.629 | Rice | e2p2 | 11 733 | -0.030 | 0.808 | -0.903 | 4.940 |
| e3p3 | 10 015 | -0.340 | 1.112 | -4.976 | 4.999 | Seafood | e3p3 | 11 212 | -0.794 | 0.631 | -4.938 | 4.900 |
| e4p4 | 2 775 | -2.045 | 1.424 | -4.992 | 4.853 | Meat | e4p4 | 7 984 | -1.274 | 1.057 | -4.994 | 4.957 |
| e5p5 | 8 291 | -0.635 | 1.614 | -4.988 | 4.992 | Eggs/Milk | e5p5 | 11 036 | 0.052 | 1.453 | -4.999 | 5.000 |
| e6p6 | 11 812 | -0.963 | 0.309 | -3.645 | 4.553 | Vegetables | e6p6 | 11 889 | -1.048 | 0.268 | -4.802 | 3.350 |
| e7p7 | 8 978 | -1.128 | 0.516 | -4.994 | 0.100 | Soybeans/Nuts | e7p7 | 10 213 | -0.633 | 0.856 | -3.968 | 4.978 |
| e8p8 | 6 993 | -1.266 | 1.000 | -4.989 | 4.879 | Fruits | e8p8 | 10 729 | -0.301 | 1.051 | -4.996 | 4.976 |
| e9p9 | 11 394 | -0.888 | 0.863 | -4.985 | 4.991 | Oil/Fat | e9p9 | 11 875 | -0.785 | 0.461 | -4.780 | 4.784 |
| e10p10 | 11 891 | -1.018 | 1.161 | -5.000 | 4.930 | Ready-made | e10p10 | 12 293 | -0.682 | 0.705 | -4.852 | 4.940 |
| e11p11 | 11 898 | -1.076 | 0.748 | -4.882 | 4.975 | Seasoning | e11p11 | 11 824 | -1.073 | 0.871 | -4.998 | 4.987 |
| e12p12 | 8 911 | -1.117 | 0.667 | -4.987 | 2.364 | Non food | e12p12 | 10 033 | -0.596 | 0.796 | -4.824 | 4.950 |
| e13p13 | 12 425 | -1.378 | 0.615 | -4.939 | 4.957 | Other food | e13p13 | 12 437 | -0.987 | 0.509 | -4.302 | 3.514 |

| 40-60 | | | | | | | Total | | | | | |
|---|---|---|---|---|---|---|---|---|---|---|---|---|
| Variable | Obs | Mean | Std. Dev. | Min | Max | | Variable | Obs | Mean | Std. Dev. | Min | Max |
| e1p1 | 3 479 | -2.789 | 1.449 | -4.998 | 4.956 | Grains | e1p1 | 16 771 | -2.216 | 2.092 | -5.000 | 4.996 |
| e2p2 | 11 647 | -0.147 | 0.523 | -0.835 | 4.973 | Rice | e2p2 | 57 729 | -0.164 | 0.594 | -0.908 | 4.973 |
| e3p3 | 10 600 | -0.820 | 0.647 | -4.980 | 4.959 | Seafood | e3p3 | 51 496 | -0.621 | 0.847 | -4.980 | 4.999 |
| e4p4 | 4 057 | -1.855 | 1.264 | -4.999 | 4.945 | Meat | e4p4 | 21 769 | -1.647 | 1.223 | -4.999 | 4.980 |
| e5p5 | 8 661 | -1.282 | 1.827 | -4.999 | 4.980 | Eggs/Milk | e5p5 | 45 020 | -0.846 | 1.553 | -4.999 | 5.000 |
| e6p6 | 11 871 | -1.085 | 0.316 | -4.972 | 3.889 | Vegetables | e6p6 | 58 828 | -0.997 | 0.353 | -4.972 | 4.899 |
| e7p7 | 9 337 | -1.151 | 0.417 | -4.985 | 4.785 | Soybeans/Nuts | e7p7 | 46 666 | -0.974 | 0.623 | -4.994 | 4.978 |
| e8p8 | 8 122 | -0.567 | 0.809 | -4.830 | 4.965 | Fruits | e8p8 | 40 427 | -0.764 | 0.996 | -4.996 | 4.976 |
| e9p9 | 11 616 | -1.108 | 0.496 | -4.971 | 3.246 | Oil/Fat | e9p9 | 57 584 | -0.906 | 0.620 | -4.985 | 4.991 |
| e10p10 | 12 047 | -1.197 | 0.994 | -4.993 | 4.991 | Ready-made | e10p10 | 59 728 | -1.063 | 0.955 | -5.000 | 4.991 |
| e11p11 | 11 763 | -0.974 | 1.023 | -4.991 | 4.953 | Seasoning | e11p11 | 58 877 | -0.981 | 0.884 | -4.998 | 4.994 |
| e12p12 | 9 544 | -0.884 | 0.632 | -4.970 | 4.493 | Non food | e12p12 | 45 385 | -0.862 | 0.732 | -4.987 | 4.950 |
| e13p13 | 12 429 | -1.316 | 0.559 | -4.994 | 2.964 | Other food | e13p13 | 61 874 | -1.169 | 0.652 | -4.994 | 4.980 |

**Average elasticities**

| | 0-20 | 20-40 | 40-60 | 60-80 | 80-100 | |
|---|---|---|---|---|---|---|
| e1p1 | 0.41 | -2.37 | -2.79 | -2.70 | -2.89 | Grains |
| e2p2 | -0.39 | -0.33 | -0.15 | 0.06 | -0.03 | Rice |
| e3p3 | -0.36 | -0.34 | -0.82 | -0.71 | -0.79 | Seafood |
| e4p4 | -1.86 | -2.04 | -1.85 | -1.78 | -1.27 | Meat |
| e5p5 | -1.60 | -0.64 | -1.28 | -1.14 | 0.05 | Eggs/Milk |
| e6p6 | -0.76 | -0.96 | -1.09 | -1.12 | -1.05 | Vegetables |
| e7p7 | -1.06 | -1.13 | -1.15 | -0.94 | -0.63 | Soybeans/Nuts |
| e8p8 | -1.43 | -1.27 | -0.57 | -0.72 | -0.30 | Fruits |
| e9p9 | -0.85 | -0.89 | -1.11 | -0.90 | -0.79 | Oil/Fat |
| e10p10 | -1.27 | -1.02 | -1.20 | -1.17 | -0.68 | Ready-made |
| e11p11 | -1.06 | -1.08 | -0.97 | -0.73 | -1.07 | Seasoning |
| e12p12 | -1.16 | -1.12 | -0.88 | -0.69 | -0.60 | Non food |
| e13p13 | -0.96 | -1.38 | -1.32 | -1.21 | -0.99 | Other food |

**Table 3A.2. Expenditure elasticities of demand for Indonesia in 2010**

| 0-20 | | | | | | | 60-80 | | | | | |
|---|---|---|---|---|---|---|---|---|---|---|---|---|
| Variable | Obs | Mean | Std. Dev. | Min | Max | | Variable | Obs | Mean | Std. Dev. | Min | Max |
| e1m | 4 692 | 1.525 | 0.720 | -2.865 | 4.981 | Grains | e1m | 6 652 | 0.235 | 1.102 | -4.986 | 4.764 |
| e2m | 11 067 | 0.085 | 0.533 | -4.895 | 0.807 | Rice | e2m | 11 626 | -0.269 | 0.753 | -4.949 | 0.877 |
| e3m | 8 788 | 0.965 | 0.192 | -3.584 | 4.728 | Seafood | e3m | 10 993 | 0.668 | 0.385 | -4.878 | 0.978 |
| e4m | 1 572 | 1.039 | 0.806 | -3.327 | 4.930 | Meat | e4m | 5 558 | 1.045 | 0.365 | -3.030 | 4.634 |
| e5m | 6 957 | 0.854 | 0.385 | -3.768 | 4.691 | Eggs/Milk | e5m | 10 852 | 0.664 | 0.336 | -3.495 | 0.998 |
| e6m | 11 365 | 0.795 | 0.224 | -4.214 | 0.984 | Vegetables | e6m | 11 880 | 0.513 | 0.409 | -4.803 | 0.976 |
| e7m | 8 564 | 0.743 | 0.340 | -4.181 | 2.225 | Soybeans/Nuts | e7m | 9 755 | 0.610 | 0.472 | -4.900 | 1.434 |
| e8m | 5 263 | 1.170 | 0.625 | -3.435 | 4.822 | Fruits | e8m | 9 437 | 0.769 | 0.438 | -4.887 | 1.604 |
| e9m | 11 041 | 0.762 | 0.180 | -4.598 | 1.101 | Oil/Fat | e9m | 11 855 | 0.410 | 0.400 | -4.977 | 0.963 |
| e10m | 11 759 | 1.322 | 0.285 | 0.968 | 4.827 | Ready-made | e10m | 12 401 | 1.005 | 0.101 | -3.241 | 4.266 |
| e11m | 11 707 | 0.752 | 0.329 | -4.958 | 0.995 | Seasoning | e11m | 11 976 | 0.482 | 0.553 | -4.844 | 0.985 |
| e12m | 7 272 | 1.048 | 0.375 | -1.897 | 3.017 | Non food | e12m | 10 199 | 0.774 | 0.283 | -2.052 | 1.932 |
| e13m | 12 437 | 1.313 | 0.177 | -0.142 | 2.571 | Other food | e13m | 12 438 | 1.329 | 0.123 | 0.899 | 2.858 |
| **20-40** | | | | | | | **80-100** | | | | | |
| Variable | Obs | Mean | Std. Dev. | Min | Max | | Variable | Obs | Mean | Std. Dev. | Min | Max |
| e1m | 5 592 | 0.957 | 1.072 | -4.915 | 4.954 | Grains | e1m | 7 607 | 0.348 | 0.736 | -4.839 | 4.013 |
| e2m | 11 596 | 0.065 | 0.557 | -4.727 | 0.829 | Rice | e2m | 11 713 | -0.229 | 0.852 | -4.973 | 0.881 |
| e3m | 10 155 | 0.918 | 0.208 | -3.324 | 3.878 | Seafood | e3m | 11 221 | 0.433 | 0.565 | -4.996 | 0.963 |
| e4m | 2 908 | 0.947 | 0.567 | -4.098 | 4.924 | Meat | e4m | 8 094 | 0.725 | 0.317 | -3.532 | 2.939 |
| e5m | 9 059 | 1.085 | 0.529 | -4.193 | 4.970 | Eggs/Milk | e5m | 11 597 | 0.393 | 0.678 | -4.866 | 0.984 |
| e6m | 11 813 | 0.692 | 0.284 | -3.741 | 0.972 | Vegetables | e6m | 11 833 | 0.274 | 0.599 | -4.980 | 0.926 |
| e7m | 9 105 | 0.711 | 0.386 | -3.997 | 2.537 | Soybeans/Nuts | e7m | 10 234 | 0.325 | 0.691 | -5.000 | 2.652 |
| e8m | 7 133 | 0.808 | 0.395 | -4.175 | 4.757 | Fruits | e8m | 10 899 | 0.380 | 0.611 | -4.939 | 1.744 |
| e9m | 11 596 | 0.641 | 0.236 | -4.370 | 0.979 | Oil/Fat | e9m | 11 894 | 0.213 | 0.605 | -4.821 | 0.972 |
| e10m | 12 261 | 1.191 | 0.203 | 0.454 | 4.871 | Ready-made | e10m | 12 410 | 0.820 | 0.236 | -4.142 | 2.499 |
| e11m | 11 988 | 0.718 | 0.343 | -4.379 | 0.993 | Seasoning | e11m | 11 885 | 0.241 | 0.721 | -4.993 | 0.957 |
| e12m | 9 153 | 0.824 | 0.332 | -4.487 | 1.662 | Non food | e12m | 10 564 | 0.558 | 0.513 | -4.999 | 3.381 |
| e13m | 12 438 | 1.323 | 0.151 | 0.706 | 2.856 | Other food | e13m | 12 438 | 1.269 | 0.098 | 0.977 | 3.289 |
| **40-60** | | | | | | | **Total** | | | | | |
| Variable | Obs | Mean | Std. Dev. | Min | Max | | Variable | Obs | Mean | Std. Dev. | Min | Max |
| e1m | 5 951 | 0.575 | 1.336 | -4.945 | 4.984 | Grains | e1m | 30 494 | 0.661 | 1.111 | -4.986 | 4.984 |
| e2m | 11 611 | -0.075 | 0.661 | -4.943 | 0.851 | Rice | e2m | 57 613 | -0.087 | 0.699 | -4.973 | 0.881 |
| e3m | 10 631 | 0.823 | 0.243 | -3.362 | 1.206 | Seafood | e3m | 51 788 | 0.748 | 0.406 | -4.996 | 4.728 |
| e4m | 4 173 | 0.937 | 0.395 | -3.673 | 4.863 | Meat | e4m | 22 305 | 0.895 | 0.452 | -4.098 | 4.930 |
| e5m | 10 158 | 0.950 | 0.391 | -4.675 | 4.591 | Eggs/Milk | e5m | 48 623 | 0.765 | 0.550 | -4.866 | 4.970 |
| e6m | 11 870 | 0.615 | 0.343 | -4.754 | 0.954 | Vegetables | e6m | 58 761 | 0.576 | 0.433 | -4.980 | 0.984 |
| e7m | 9 414 | 0.679 | 0.383 | -4.527 | 2.032 | Soybeans/Nuts | e7m | 47 072 | 0.606 | 0.503 | -5.000 | 2.652 |
| e8m | 8 184 | 0.744 | 0.396 | -3.801 | 2.080 | Fruits | e8m | 40 916 | 0.719 | 0.558 | -4.939 | 4.822 |
| e9m | 11 718 | 0.546 | 0.295 | -4.028 | 0.971 | Oil/Fat | e9m | 58 104 | 0.510 | 0.422 | -4.977 | 1.101 |
| e10m | 12 359 | 1.082 | 0.104 | -0.012 | 3.847 | Ready-made | e10m | 61 190 | 1.081 | 0.261 | -4.142 | 4.871 |
| e11m | 11 965 | 0.610 | 0.439 | -4.801 | 0.980 | Seasoning | e11m | 59 521 | 0.560 | 0.532 | -4.993 | 0.995 |
| e12m | 9 784 | 0.742 | 0.255 | -1.926 | 1.619 | Non food | e12m | 46 972 | 0.771 | 0.395 | -4.999 | 3.381 |
| e13m | 12 437 | 1.334 | 0.137 | 0.887 | 2.693 | Other food | e13m | 62 188 | 1.314 | 0.142 | -0.142 | 3.289 |

**Table 3A.2. Expenditure elasticities of demand for Indonesia in 2010 (*cont.*)**

| | Average elasticities | | | | | |
|---|---|---|---|---|---|---|
| | 0-20 | 20-40 | 40-60 | 60-80 | 80-100 | |
| e1m | 1.53 | 0.96 | 0.58 | 0.24 | 0.35 | Grains |
| e2m | 0.08 | 0.06 | -0.07 | -0.27 | -0.23 | Rice |
| e3m | 0.97 | 0.92 | 0.82 | 0.67 | 0.43 | Seafood |
| e4m | 1.04 | 0.95 | 0.94 | 1.05 | 0.73 | Meat |
| e5m | 0.85 | 1.09 | 0.95 | 0.66 | 0.39 | Eggs/Milk |
| e6m | 0.80 | 0.69 | 0.62 | 0.51 | 0.27 | Vegetables |
| e7m | 0.74 | 0.71 | 0.68 | 0.61 | 0.33 | Soybeans/Nuts |
| e8m | 1.17 | 0.81 | 0.74 | 0.77 | 0.38 | Fruits |
| e9m | 0.76 | 0.64 | 0.55 | 0.41 | 0.21 | Oil/Fat |
| e10m | 1.32 | 1.19 | 1.08 | 1.00 | 0.82 | Ready-made |
| e11m | 0.75 | 0.72 | 0.61 | 0.48 | 0.24 | Seasoning |
| e12m | 1.05 | 0.82 | 0.74 | 0.77 | 0.56 | Non food |
| e13m | 1.31 | 1.32 | 1.33 | 1.33 | 1.27 | Other food |

## Annex 3.B

## The general equilibrium model INDONESIA-E3[1]

The general equilibrium model used in Step 2 is known as INDONESIA-E3. It belongs to the Johansen class of general equilibrium models, which are linear in percentage changes. Most structural features are as described in Warr and Yusuf (2014) to which the reader is referred. More detailed discussion of data sources used and parametric values is provided in Yusuf (2006, 2008).

*Factor mobility*

The labour force is segmented into 'skilled' and 'unskilled', based on workers' occupations. Skilled labour means clerical and managerial workers and unskilled means agricultural production workers and non-agricultural manual workers. Both categories of labour are assumed to be mobile across all sectors while capital and land are immobile across industries. These features imply an intermediate-run focus for the analysis, with an adjustment time of about two years. The focus is neither very short-run, or else labour would be less than fully mobile, nor long-run, or else capital and land would be more mobile.

*Households and final demand*

Two categories of households are identified, rural and urban, each divided into 100 sub-categories of equal population size, with these sub-categories arranged in order of expenditure per capita. Urban and rural households differ considerably, particularly as regards endowments of skilled and unskilled labour. Ownership of rural land is surprisingly important among urban households. Net transfers between households are relatively minor. Within each of the urban and rural categories, there is considerable variation in factor ownership. The principal source of the factor ownership data is Indonesia's Social Accounting Matrix for 2003, supplemented by additional information outlined in Yusuf (2006).

International price changes and other shocks will normally produce both gainers and losers. We wish to discover the *net* effects on food security. Disaggregation of the total population into its rural and urban components suits this objective and is policy-relevant. But the disaggregation might in principle have been done differently, such as division by socio-economic group or by occupational category – instead of, or in addition to, the rural/urban split employed here.

*International trade*

Indonesia is assumed to trade with the rest of the world at exogenously given international prices. On the import side, especially important for staple foods, each imported food commodity is a close but imperfect substitute for the corresponding domestically produced food item.[2] The degree of substitutability is captured by the Armington elasticity of substitution and the values of this parameter were obtained from the Global Trade Analysis Project (GTAP) database.[3] These values are: soybeans 2.25, wheat 2 and maize 1.3. The Armington elasticity for rice was set at 20, a higher value than the GTAP estimate, reflecting the assumption that Indonesian rice and imported rice (now primarily from Vietnam) are very close substitutes. On the export side, the supply of each exported commodity is modelled with a constant elasticity of transformation between production for exports and domestic uses.

*Model closure*

The macroeconomic features of the model closure and the reasons for them are as described in Warr and Yusuf (2014). Transfers received by households are exogenous, but all components of household factor incomes and all consumer prices are endogenous. An important feature of the model closure relates to the treatment of rice imports. As described below, since 2004 Indonesia has officially banned rice imports above a minimal level. The model closure reflects this fact by specifying the level of rice imports exogenously and allowing the domestic price of rice to be determined endogenously. The difference between the domestic wholesale price of rice and the *c.i.f.* import price thus constitutes a rent accruing to import license holders, assumed to be the richest five per cent of urban households.

*Shocks: Simulation outputs*

For each of the simulations of an exogenous shock, the general equilibrium modelling output has two components. The first is a set of changes in commodity prices, aggregated to the categories used in the estimated demand system. Because of the focus on food security, this commodity aggregation focuses on food and distinguishes separately rice, grains & tubers, seafood, meats, eggs & milk, vegetables, soybeans & nuts, fruits, oils & fats, delicatessen items & drinks, seasoning (mostly chili pepper, salt, sugar), other foods, and one non-food aggregate

The second component of the output is the changes in the nominal expenditures of each of the 200 households (100 urban and 100 rural) identified in the model. These two sets of information are then used to estimate changes in food security.

## *Notes*

1. This material, and details of the use of INDONESIA-E3 elsewhere in the report, is based on a consultant paper by Peter Warr from the Australian National University.
2. Wheat is an exception in that wheat is not produced within Indonesia at all.
3. See https://www.gtap.agecon.purdue.edu/

# Annex 3.C

## Additional risk scenarios

A total of ten scenarios were identified and quantified by the consultative process, but only five of them plus the reference scenario were retained for full policy analysis and in-depth discussion with policy makers and stakeholders. In this annex, details are given of the remaining five scenarios.

*Scenario 6: Increase in international soybean prices*

As with rice, Indonesia is a large net importer of soybeans. But whereas rice contributes mainly to energy requirements, soybeans are especially important for protein intake. Soybeans are used in manufacturing a wide variety of processed foods consumed by humans and for animal feed, producing protein-rich foods based on livestock, especially poultry products. But soybeans are not consumed directly by humans. Because Indonesia's soybean production is not large and is undertaken by a relatively small number of producers, very few Indonesian households benefit from an increase in soybean prices. Rates of protection of soybeans within Indonesia have been low and international price changes have tended to be transmitted directly to domestic wholesale prices (Fane and Warr, 2009).

The recent history of soybeans policy in Indonesia is that until 1996 the government protected soybean growers by assigning BULOG a monopoly on imports. Since 1996, soybean imports have been unrestricted, and the tariff is currently zero. For some years prior to that date, a local soybean crushing plant run by *PT Sarpindo Industri* was protected by a local-content scheme that required domestic feed mills to source at least 20% of their total usage of soybean meal from local supplies, which benefited *Sarpindo*, the only local supplier. The high cost of feed inhibited the growth of the increasingly powerful poultry industry. In 1996, the local-content scheme was abandoned, and *Sarpindo* was allowed to go out of business (Fane and Warr, 2009).

Since around 2000 soybean imports have been freely permitted, but it is claimed that soybean processing has become oligopolistic. Although these claims are largely unproven, the implication would be a rise in domestic soybean prices relative to world prices, which has actually occurred, and a decline in price transmission from the international price to the domestic market price.

Figure 4A.1 shows that the international price of soybeans has been less volatile than the international price of rice. Sharp increases occurred in 1973-74, 2002 and especially 2007-08, when the US dollar price increased by 118% in one year. During the Asian Financial Crisis of 1998, the value of the Indonesian rupiah collapsed and domestic prices, measured in rupiah, surged but not by as much as the exchange rate-adjusted international price. Domestic prices declined, measured in dollars. Since the crisis of 1998, Indonesian domestic prices have followed international prices, measured in US dollars, except that the domestic price increases since 2008 have exceeded the rise in the international price, due to a supply-induced shortage of soybeans. Further shocks of that kind are very possible.

Given the policy setting, an increase in the international price is expected to be transmitted to domestic soybeans prices and thereby raise the price of soybean meal, increasing production costs in the important poultry industry. A cost increase in poultry production would be transmitted to poultry meat prices, resulting in lower poultry meat consumption and hence reduced protein intake.

Like scenario 1, this scenario is also based on the 2007-08 international price shock. It assumes an international real price increase of soybeans, holding all other commodity prices constant, of 118%, again replicating the actual increase that occurred in 2007-08. The likelihood of an international price shock of this magnitude, based on past price behaviour, is about once every 30 years. It is expected that the internal price of soybeans would increase and that the prevalence of undernourishment would increase, along with under-consumption of protein.

**Figure 3.C1. Indonesia: Wholesale prices and world prices for soybeans, 1961 to 2013**

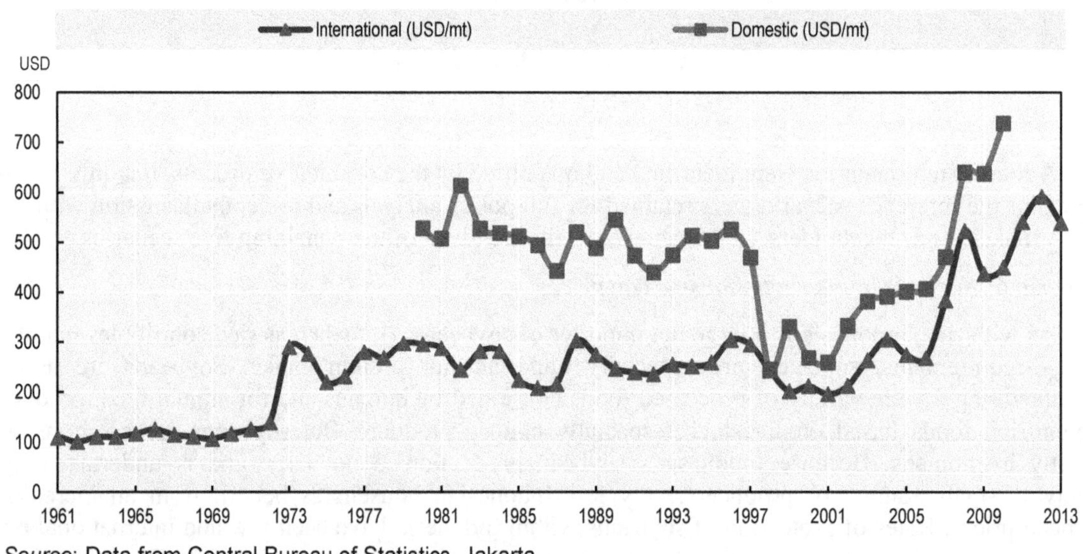


*Source*: Data from Central Bureau of Statistics, Jakarta.

*Scenario 7: Increased world prices: rice, soybeans, wheat and maize*

**Figure 3.C2. International nominal prices for rice, soybeans, wheat and maize 1990 to 2012**

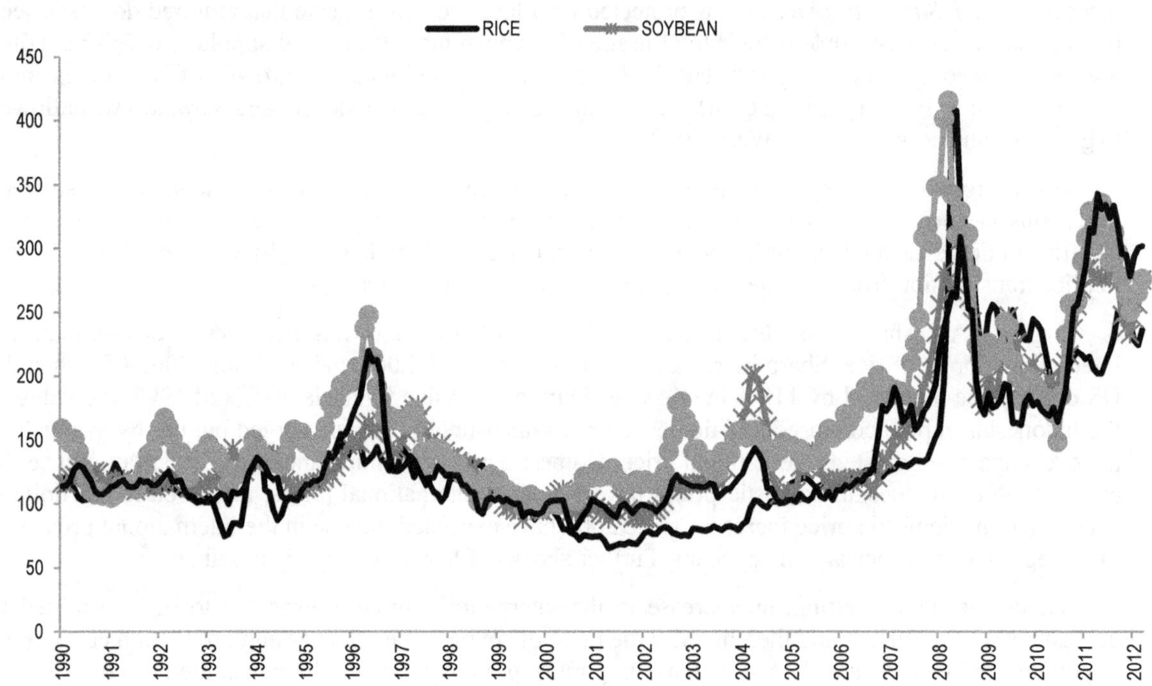


*Sources*: Data from International Financial Statistics (http://www.imfstatistics.org/imf/), except maize for May 2011 onwards, for which data are from http://ycharts.com/indicators/us_maize_price_gulf_ports.

**Table 3.C1. International price changes, Indonesia's food imports**

(Jan-June 2003 to Jan.-June 2008 – per cent)

| | Rice | Soybeans | Wheat | Maize |
|---|---|---|---|---|
| Nominal price change | 287 | 169 | 251 | 178 |
| Real price change, using MUV Index as deflator | 212 | 117 | 183 | 124 |

*Source*: Calculations by Peter Warr using data from International Financial Statistics (http://www.imfstatistics.org/imf/), except maize for May 2011 onwards, for which data are from http://ycharts.com/indicators/us_maize_price_gulf_ports. MUV index from World Bank.

In the case of the 2007-08 food price crisis, the international real prices of rice, wheat, maize and soybeans all increased markedly. These are all important import commodities for Indonesia. The nominal prices are summarised in Figure 4A.2 and their real price implications, using the World Bank's Manufacturing Unit Value Index as deflator, are summarised in Table 3.C1.

This scenario examines the implications of increases in these real prices at the rates seen in the 2007-08 price shock. Indonesian trade policy is assumed to be as in Scenario 1-a for rice and as in scenario 6 for soybeans. The likelihood of a combination of international price shocks of this magnitude, based on past price behaviour, is about once every 40 years. It is expected that under this scenario food insecurity would be worsened, but only moderately.

*Scenario 8: International financial crisis*

The Asian Financial Crisis (AFC) of 1997-99 had devastating effects on Indonesian households, including their food security. The international value of the Indonesian rupiah collapsed in January 1998, losing more than two-third of its value against the US dollar in a few days (Johnson, 1998). Per capita real GDP fell by 11.4% between 1998 and 1999. The annual inflation rate rose to 80% in 1998 (Figure 3.C3), while food prices, especially of staples, increased 20% more than the general consumer price index. Poverty incidence temporarily doubled, from 11.3% to 23.8% of the population (Hill, 2012).

Based on the Indonesia Family Life Survey, Thomas and Frankenberg (2007) estimated that the 1997-98 financial crisis led to a decline of real per capita expenditure for urban and rural household by 34% and 18% on average, respectively. They also estimated that the share of food expenditure increased on average by 8% for urban households and 6% for rural households.

**Figure 3.C3. Real GDP and inflation in Indonesia, 1970 to 2013**

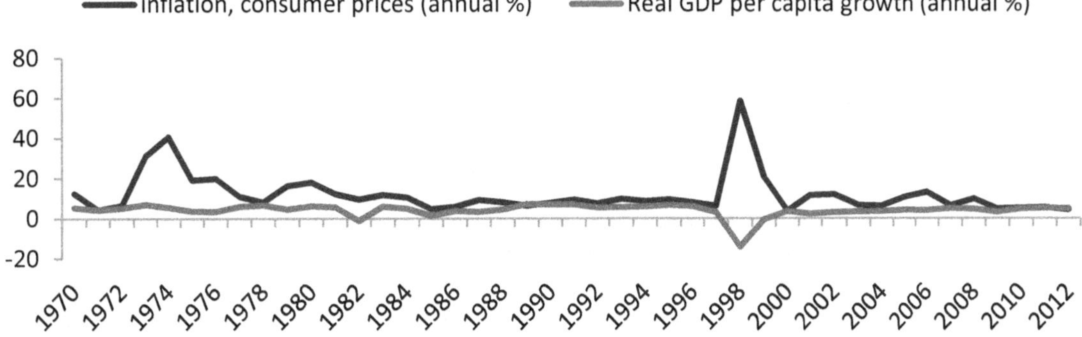


*Source*: World Bank, World Development Indicators.

An international financial crisis does not necessarily lead to a macroeconomic crisis within Indonesia. A good example is the Global Financial Crisis of 2008-2010, which produced virtually no contraction within Indonesia. Indonesian financial markets were not significantly affected. The impact on Asian developing countries was primarily through reduced demand for their exports. Measured as a share of GDP, Indonesia's export dependence is relatively low by Asian standards – much lower than Thailand or Malaysia, for example – and the effect of the global slowdown was correspondingly small.

A financial crisis, however caused, would lower investor confidence, producing a contraction of output and employment in industries particularly sensitive to changes in aggregate investment. The Asian Financial Crisis caused a 40% contraction of private employment in the formal sector (Johnson, 1998). The construction industry was the most heavily affected. The incomes of construction workers fell, along with remittances to their families elsewhere in Indonesia (Pardede, 2006). An international financial crisis would also affect the incomes of Indonesians working abroad and the remittances they send to their families in Indonesia. Because large numbers of Indonesians are found at income levels just above the poverty line, these reductions in their incomes can be expected to have large effects on poverty incidence and the prevalence of food insecurity.

Special circumstances caused the AFC: high volumes of internationally mobile capital relative to Indonesia's international reserves and, most especially, a policy commitment to maintaining a fixed exchange rate, circumstances that no longer exist. An international financial crisis is possible, but it would not take the same form as the AFC of the late 1990s and would almost certainly be less severe.

Scenario 8 analyses the effect of a 10% reduction in investment demand in Indonesia, accompanied by a 10% reduction in export demand. Although agricultural output is not affected directly, a decline in household incomes is expected, inducing a fall in food demand and a consequent worsening of food security. The likelihood of an international shock of this magnitude, based on past events, is about once every 30 years. The prevalence of undernourishment is expected to increase moderately.

*Scenario 9: Systemic drought in Java due to El Niño*

Indonesia's climate is influenced by the *El Niño*/Southern Oscillation (ENSO) events. About 93% of droughts recorded in Indonesia have occurred during *El Niño* years. ENSO events occur typically on a three- to seven-year cycle (Salafsky, 1994). Naylor et al. (2001) have noted four strong *El Niño*-induced dry periods having important consequences for rice production during the period 1970 to 2000. They provide a detailed discussion of the effects of each of these events. In years of yield decline, domestic production was complemented by imports. In 1998, crop failure allied with the AFC economic crisis led to a 300% increase in the price of food grain.

*El Niño* droughts have important effects on rice production, particularly on Java Island, where more than 50% of Indonesia's rice is grown (Figure 3.C4). Naylor et al. (2001) estimated a loss of paddy rice production in Java of 4.8 million tons during the 1997-98 droughts, equivalent to 17% of Java's annual rice production.

The economics of drought are paradoxical. When imports into Indonesia are fixed, a temporary loss of domestic production in Java drives up domestic nominal and real prices of agricultural products. Because the demand for these products is inelastic, the proportional increase in the price will exceed the proportional reduction in output, so that the total revenue received by producers actually rises.

In the case of rice, the increase in the price within Java is moderated by shipments of rice from other islands, where rice is also produced. But in the case of soybeans, where Java dominates domestic production, there is much less scope for that to happen, at least in the short run. Additional supplies can be obtained only by importing from the other side of the world. This takes time. In the meantime, soybean prices rise, with indirect effects on the prices of other foods that use soybeans as inputs.

Because of the regional nature of this scenario – its focus on the island of Java – the analysis of this scenario combines two general equilibrium models: the INDONESIA-E3 model described above, and TERM (The Enormous Regional Model), a regional model of the Indonesian economy constructed at

Monash University, Melbourne (Wittwer, 2012). INDONESIA-E3 has a multi-household structure, but no regional disaggregation. TERM has a regional disaggregation but no household disaggregation. The procedure is to use TERM to analyse the overall economic effects of the shock on Java Island and then to feed this information into INDONESIA-E3 to obtain the effects on households. The assumption being made is that the structure of effects across households within Java is approximated by the structure of these effects within Indonesia as a whole, as captured within INDONESIA-E3. This procedure was also used with scenario 5 (earthquake on Sumatra).

Scenario 9 analyses the effect of a 10% contraction in productivity in all agricultural commodities within Java. The likelihood of drought of this magnitude, based on past experience, is about once every seven years. A worsening of food security is expected, depending on the trade policy in place, but many rural households may gain from the drought, at least in the short-run, because of increased incomes from agricultural product price increases. Urban consumers, however, will be harmed by the increases in rice and soybean prices.

**Figure 3.C4. Production and yield of rice in Java**

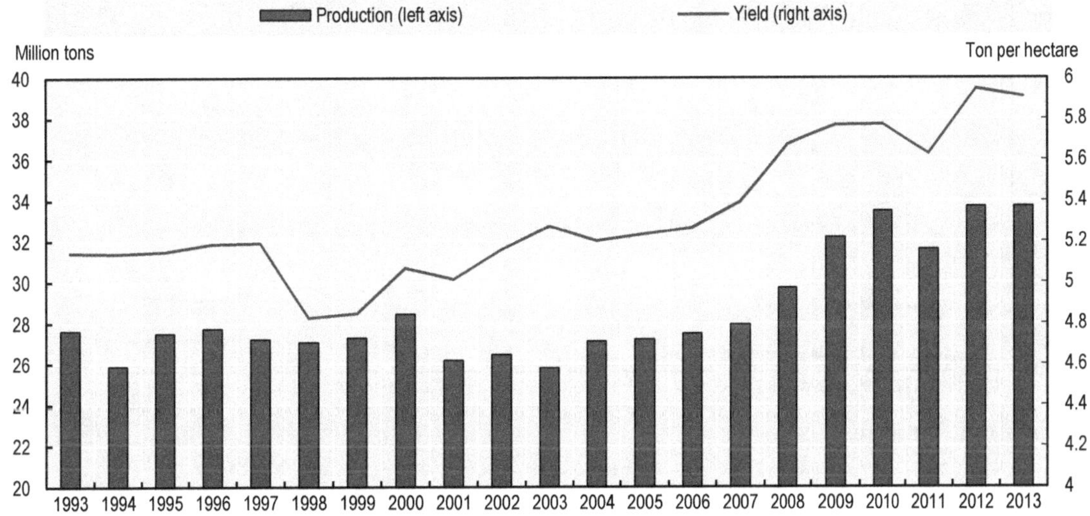


*Source*: Data from Central Bureau of Statistics, Jakarta.

*Scenario 10: Avian influenza epidemic*

In 2004, a major outbreak of the H5N1 strain of avian influenza had a significant effect on poultry production and consumption (Figure 3.C5). The outbreak carried with it the danger of transmission from poultry to humans, and by May 2006, 37 human deaths had been attributed to this virus. The effects on the poultry industry included the necessary destruction of large numbers of birds. In Central Java and Bali, nearly one quarter of all birds were culled. Fortunately, the effect was not as large in other parts of Indonesia. Of the 30 provinces of Indonesia, 15 were significantly affected. Employment in the commercial poultry industry fell by 23%. Overall, about one-fifth of total poultry numbers was lost.

The total direct economic losses were USD 171 million, estimated by Hall et al. (2006), and indirect losses estimated by the Indonesian Poultry Information Centre were USD 171 million, but neither of these estimates included losses incurred by village and backyard farmers, the largest part of the industry. Elci (2006) also reports a virtual cessation of exports of chicken from Indonesia in 2004 compared with exports around USD 1.5 million in preceding years.

Indonesia's poultry industry is vast, including both smallholder and large-scale production, and contributes a large share of Indonesian households' total animal protein intake. Data from CASERD of the Indonesian Ministry of Agriculture indicates that in 2004 Indonesia's total poultry population was

275 million birds, of which 175 million were in backyard production, especially in rural areas, and the remainder in commercial broiler and layer establishments.

Scenario 10 considers a 20% reduction in domestic poultry production, focusing on its effects on the protein intake of Indonesian households. The likelihood of a negative shock to poultry productivity of this magnitude, based on past experience, is about once every 15 years. It is expected that farmers' incomes will be reduced, damaging their food security, and that protein intake within Indonesia will also be lower as poultry product prices rise in response to the contraction in supply. Calorie intake may even increase as shortage of poultry products induces increased consumption of calorie-intensive staples.

**Figure 3.C5. Poultry numbers in Indonesia (millions of birds)**

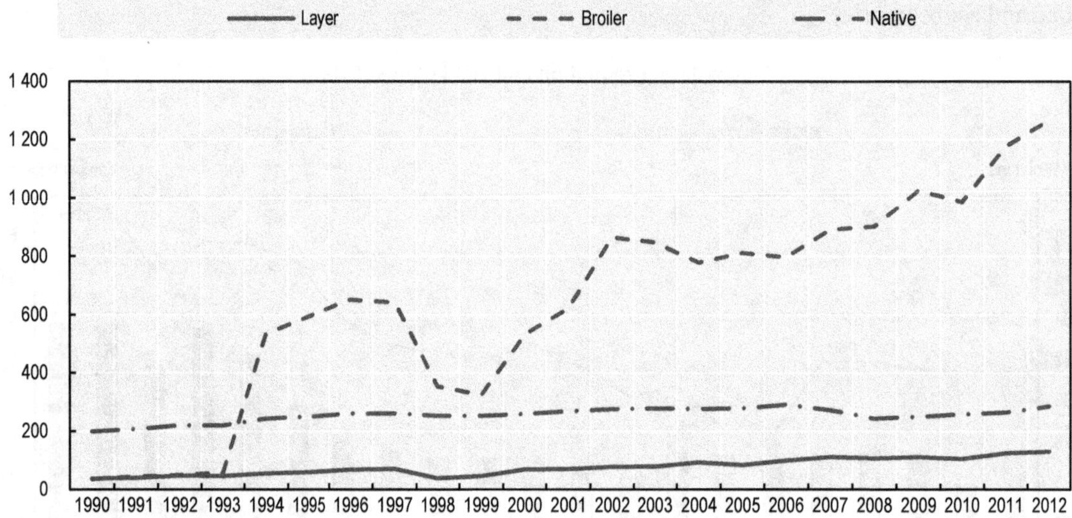


*Source*: CEIC Database.

# *Chapter 4*

# Policy discussion, conclusions and recommendations

*This chapter summarises policy lessons from the OECD Framework for the Analysis of Transitory Food Insecurity and its application to Indonesia. It contains six specific policy recommendations to improve policy approaches to food insecurity risks in Indonesia.*

This study applies the OECD Framework for the Analysis of Transitory Food Insecurity to Indonesia and the risk management portfolio approach to a set of policies that have differentiated impacts under different contingent scenarios. Potential food insecurity scenarios were identified through consultation with the stakeholders in Indonesia. The study finds that domestic economic and natural disaster scenarios are more important than global price hikes both in terms of their likelihood and their potential impact on food insecurity. This result should guide policy design. A set of existing and alternative agriculture and social policy options have been analysed: rice price support measures, the rice consumer subsidies programme Raskin, a food voucher programme, the social programme of unconditional cash transfers BLT, the fertiliser subsidy programme and a crop insurance programme. The approach to policy assessment in managing transitory food insecurity risks provides policy insights about the trade-offs of different policy options.

Market price management measures that attempt to stabilise and support domestic prices can only be undertaken in a context of trade restrictive measures. The OECD estimates that the domestic rice price was 60% higher than the reference international price in 2010-12, compared to 8% in 2000-2.The set of rice price support measures examined here (domestic public purchase plus import and export restrictions) does not contribute to improve any dimension of food security including stability, but instead worsens the situation (increasing undernourishment by between 2 and 22 percentage points of the population). Export restrictions can help avoid a surge of undernourishment only in the case of a price spike, as happened in 2008, which is expected to occur only once in 30 years. The price support that has remained after 2008 worsens the food security situation in all other scenarios. At the same time, export restrictions by all countries would be expected to directly exacerbate the price spike and its negative consequences for the poor (FAO et al., 2011). All countries following self-sufficiency and trade restrictive policies reduce the risk pool and the transfer of risk across countries, which increases the risk of international price spikes. There are additional long-term effects of constraining exports, which in turn discourages investment and growth in output. Facilitating imports and the active participation of Indonesian and foreign traders and investors can contribute to rural growth, incomes, and food supplies. The current focus on self-sufficiency and restrictive trade measures is influenced by political economy factors. It may also in part be the result of overestimating the importance of the price hike scenario in relation to normal market conditions and other relevant scenarios, and of underestimating the likelihood of domestic natural disasters that require quick access to imports to ensure food security. The government should develop a portfolio of policies that can more robustly respond to a diversity of food insecurity scenarios.

From a risk management perspective, it is not efficient to use price support measures that induce food insecurity in most scenarios, can only improve food security in a single unlikely scenario, and increase the variability of food security outcomes. It is preferable to consider other ways of dealing with the specificities of the extreme scenario of a rice price spike. The ASEAN Economic Community 2015 could be a good opportunity to launch a new regional trade-oriented approach, including a coordination agreement within ASEAN to restrain the use of export restrictions and eliminate the administrative requirement of export permits. Export restrictions are very damaging for global and regional food security and, when prices increase, they can create policy traps that can exacerbate lead to price spikes. The ASEAN region has the world's two largest exporters and importers of rice, and they strongly influence what happens on international rice markets. Freer and more predictable regional trade with a mutual commitment not to restrict imports or exports and to eliminate administrative trade permits can ensure the regional availability of rice and reduce its price variability. Further analysis would be needed to quantify these opportunities. A complementary initiative would be the development of reliable futures markets for rice, so that both the private sector and the governments can buy protection against large price movements.

The existing Raskin and unconditional cash transfers BLT assistance programmes are more robust across scenarios, despite their limited degree of targeting to the poor and to the undernourished. Their impact on each scenario is not large (around one per cent of the population is brought above the undernourishment threshold), but it is always positive. Furthermore, the BLT programme significantly reduces the variability of the impacts on undernourishment.

Because of the logistical costs of managing physical rice, Raskin has very large transaction costs and is less cost-efficient than BLT. The Raskin programme concerns only rice, which is the primary source of calories, and it has a more significant effect on calorie intake than do cash transfers. However, this single focus reduces the dietary diversification of poor people, which is another nutritional policy objective in Indonesia. Moreover, rice is one of the most price-inelastic commodities in Indonesia, which means that a large price subsidy has only a modest impact on calorie intake. Experience in other countries has shown that, in the long run, maintaining consumer subsidies with high producer prices may become unsustainable because price gaps become too expensive to finance (e.g. Mexico, see OECD, 2006). Raskin and price support measures are not complementary from a risk management perspective: Raskin is clearly unable to offset the negative impacts of price support on undernourishment, and preliminary results show that an easy way to improve food security across the risk scenarios is to reduce the current rice price gap.

This study has also investigated the potential of fertiliser subsidies and a hypothetical crop insurance programme for improving food security across all scenarios. The impact of the former is very small regardless of the scenario. The impact of the latter is only slightly bigger in the crop failure and earthquake scenarios and zero in other scenarios. These programmes are focused on the farming population, but their degree of targeting to the most vulnerable farmers is not obvious. The main advantages of these programmes for food security are their potential for increasing food availability and, in the case of insurance, for mitigating farmers' income losses after a crop failure. However, the literature shows that fertiliser subsidies have low transfer efficiency and perform poorly with respect to sustainably increasing agricultural growth (Jayne and Rashid, 2013).

In the case of insurance, two different objectives may be pursued: the protection of vulnerable farm households against production risks, and the development of risk management tools that can enhance investment in the sector. For the first objective, it is difficult and expensive to make insurance available to poor farmers who would benefit more from social programmes; these programmes could add farming specificities such as yield or weather triggers to adjust payments. The main potential for insurance is to improve agricultural investment by making risk management tools available to individuals or groups of farmers. For this purpose, the involvement of professional insurers with know-how beyond the state or local owned enterprises is crucial.

Further research should focus on ways of improving social assistance programmes. Several avenues could be pursued, in particular the food vouchers and targeted cash transfers explored in this study. Raskin could be converted into vouchers usable to buy a range of different basic food items that could be differentiated regionally to respond well to local food preferences. This would allow a broader focus on staple food consumption and undernourishment objectives by giving poor households the freedom to buy other food staples if rice becomes too expensive or demand is too inelastic. It would also reduce substantially the logistical costs of physically buying and delivering rice, activities that the private sector could undertake more efficiently. This movement would decouple consumer assistance from the maintenance of a rice price gap, eliminating the mutual reinforcement between these two sets of policies managed by the same agency BULOG. At the same time, BULOG should phase out its domestic market interventions over time, re-focussing on the neutral management of emergency food reserves. This movement towards domestic policy reform could well be a first step that facilitates subsequent reductions in trade restrictions. This sequencing of domestic and trade policy reform has facilitated the political economy of reform in other countries[1]. Further analysis should be undertaken to define a good governance structure for the emergency reserve system and the links with the sub-national reserves and ASEAN+3 emergency rice reserve system (APTERR).

BLT gives cash transfers which can be spent freely on food and non-food items. The result is a more effective reduction in poverty than in undernourishment. There is clearly scope to improve the degree of targeting of social programmes such as BTL. Better targeting could mean both improvements in identifying the recipients and in attaching conditions. In June 2013, the Indonesian government issued Social Protection Cards (KPS). The income situation of the cardholders can be followed, matching the information on the card about social programme benefits with expenditure information regularly collected from other sources, such as for the large SUSENAS survey. The identification cards for social programmes can be used to improve targeting over time. The amount of the payment could also be modulated according to some indicator or trigger. The payments could be conditional on food expenditure or school enrolment such as food voucher or school lunch coupon. Such fine-tuning of these measures could significantly improve their effectiveness in managing the instability of food security. The Social Protection Card is a good first step towards the convergence of social programmes that could include specificities for farmers based, for instance, on weather indexes. The convergence of social and food aid programmes should continue and the proposed food voucher programme should be jointly managed with other social programmes to improve effectiveness and enable better monitoring of results.

There is scope for transforming fertiliser subsidies into more effective policy measures. In particular, these significant budgetary resources could be channelled for strategic public investment in a resilient food and agriculture system in Indonesia, including investment in people (education, training, extension services) and in physical infrastructure. An interesting long-term investment for Indonesia could be developing a good weather information network, with a dense map of weather stations and an information tool that could be used by farmers to assess and manage their risks, by insurers or financial institutions as tools to develop more sophisticated financial instruments like index insurance, and by governments to guide decisions on disaster assistance. The investment in developing high-yield and pest resilient rice varieties could also be rewarding. Further investigation would be needed to better define the orientation of these long-term investment policies, and priorities should be identified in consultation with regional groups.

This report presents a method for organising the impacts of policies addressing transitory food insecurity in a simple matrix of policy measures and scenarios, which can be a powerful instrument for understanding the trade-offs between different policy options in the area of transitory food insecurity. Better targeting of policy can significantly improve the stability dimension of food security in Indonesia.

The usefulness of this policy analysis would be enhanced by a broader analysis of other objectives, in particular long-term objectives related to development and poverty reduction. However, this basic portfolio analysis considers only a limited number of diverse scenarios, and a limited number of possible policy combinations. Important aspects of food security associated with the development of social capital and logistic infrastructure, particularly in rural and isolated communities, are not part of the analysis but are an essential complement of the policy recommendations.

These limitations should not overshadow the strengths of this analysis, that is, its simplicity, rigour and responsiveness to the risk perceptions of experts and policy makers. In particular, several general insights and some specific policy recommendations emerge from the analysis. The general insights, while anchored in the Indonesian context, are nevertheless pertinent to other countries tackling the issue of policies that are conducive to risk management.

First, there is scope for benefits from policy diversification. For instance, certain sanitary and phytosanitary measures can be very effective in scenarios involving food safety risk. Although their impact in other scenarios, such as an earthquake, are probably zero, measures to protect and enhance animal and plant health are, to a certain extent, of net benefit because they mitigate the risks of some scenarios such as a food safety crisis without causing harm in other scenarios. The more diverse the impact of different policy instruments in different scenarios, the larger the scope for policy diversification. A policy strategy that concentrates on addressing a single source of risk, such as price spike in international markets, would increase a vulnerability to other sources of risks such as domestic crop failure.

Second, and as a qualification to the previous observation, some policies that are positive in one scenario may have negative food security impacts in other scenarios. All policies tackling transitory threats to food security need to be consistent with the whole policy strategy for global food security. For instance, tariff measures that increase self-sufficiency may be thought to respond to the threat of potential import disruptions, but they increase food prices in all other scenarios when the price shock is coming from domestic source. The result of these price support policies is an increase in food insecurity in most scenarios, including in the reference scenario with no particular shock. Policy makers should keep in mind the unintended impacts of a policy strategy. Thus, an effective portfolio would contain diverse policies that deal well with one or more food security risks, whilst having neutral impacts in other types of situation.

Third, the Indonesian case study shows that a key instrument for rigorous policy analysis of transitory shocks to food security is an on-going household expenditure survey, which allows the quantitative measurement of food poverty and undernourishment. Using appropriate economic and statistical methods, the impact of different risk and policy scenarios can be estimated. These estimates are critical for a rigorous risk and policy assessment.

There are six specific policy recommendations that emerge from the analysis of the risks to Indonesia's food security. The analysis suggests that it would be beneficial for the food security situation to:

- *Dismantle the rice subsidy programme Raskin and replace it with a food voucher programme* that, for the same cost, could be better targeted to the most vulnerable segment of the population. Food vouchers would be used to buy food staples consisting not only of rice but including other basic items. The exact list of products covered by the voucher should be decided in consultation with regional groups and could be differentiated regionally to respond well to local food preferences.
- *Improve the targeting of unconditional cash transfers*, for example by including triggers based on income and possibly special provisions based on weather or production losses for farmers. The convergence of social and food aid programmes should continue, and the new food voucher programme should be jointly managed with other social programmes to improve effectiveness and enable better monitoring of results.
- *Reform BULOG, by reducing its commercial activities and re-focussing its activities on the neutral management of emergency food reserves*. The floor purchasing price of rice should be phased out over time. Further analysis should be undertaken to define a good governance structure for the emergency reserve system and the links with the sub-national reserves and ASEAN+3 emergency rice reserve system (APTERR).
- *Reform the administrative requirements for agro-food imports, including import permits for rice*. Facilitating imports and the active participation of the Indonesian and foreign traders and investors can contribute to rural growth, incomes, and food supplies.
- *Promote a coordination agreement within ASEAN to restrain the use of export restrictions* and eliminate the administrative requirement of export permits. Export restrictions are very damaging for global and regional food security and, when prices increase, they can create policy traps that can exacerbate price spikes. The ASEAN region includes large exporters and importers of rice. More open and reliable regional trade among these trading countries could conceivably do more to reduce the variability of rice prices and ensure availability in all countries.
- *Phase out fertiliser subsidies and use the released budgetary amounts for strategic public investments,* including investment in people (education, training, extension services) and in physical infrastructure. Priorities should be identified in consultation with regional groups.

Building resilience to different types of food security shocks at the local, regional and national level is a complex objective that requires a coherent set of policies. The governance of the policy design, implementation and response to food security risks is an additional challenge that needs strong engagement of local and community organisations, the participation of a multidisciplinary pool of experts, and a strong coordination across ministries, agencies and levels of government. Local and community organisations often provide the first response to shocks, particularly in the case of natural disasters, and they are crucial for managing and implementing policy action. The complexities and uncertainties of the design of sound policies discussed in previous sections provide another reason for the government to invest in capacity-building in local organisations to improve their knowledge and social capital, and to enhance the participation of experts and stakeholders in the risk assessment and the design and implementation of policy responses.

## *Note*

1. Despite many differences between the two countries, a relevant example is the case of Switzerland, which has undertaken domestic reform of their direct payment system as a first step to reducing border measures that have supported domestic prices since the early 1990s. The example of CONASUPO in Mexico is probably closer to the reality of Indonesia; the elimination of Mexico's general consumption subsidy for maize in the early 1990s facilitated and reinforced the gradual reduction in border measures in Mexico, particularly towards the other NAFTA countries in the late 1990s and early 2000s.

## References

Antón, J. et al. (2013), "Agricultural Risk Management Policies Under Climate Uncertainty". *Global Environmental Change*, Vol. 23, pp. 1726–1736.

Badan Pusat Statistik, "Consumer Price Indices in 66 Cities in Indonesia (2007=100) 2008-10".

Badan Pusat Statistik, SUSENAS database 2005-10.

Baldi, G., et al. (2013), "Cost of the Diet (CoD) Tool: First Results from Indonesia and Applications for Policy Discussion on Food and Nutrition Security", *Food and Nutrition Bulletin*, Vol. 34, No. 2 (supplement), The United Nations University, Tokyo.

Banks, J., R.W. Blundell, and A. Lewbel (1997). "Quadratic Engel Curves, Welfare Measurement and Consumer Demand," *Review of' Economics and Statistics*, Vol. LXXIX(4), pp. 27-539.

Cafiero, C. (2013), "What Do We Really Know About Food Security?" *National Bureau of Economic Research Working Paper Series*, N°18861.

Cafiero, C. and P. Gennari (2011), *The FAO indicator of prevalence of undernourishment*, FAO publications, Rome.

Deaton, A. (1997), *The Analysis of Household Surveys*, World Bank and the Johns Hopkins University Press, Washington D.C.

Deaton, A. (1988), "Quality, quantity, and spatial variation of price", *American Economic Review*, Vol.78, No.3, pp.418-30

Deaton, A.S., and J. Muellbauer (1980). "An Almost Ideal Demand System," *American Economic Review,* Vol. 70, No. 3, pp. 312-326.

DEFRA (2009), *UK Food Security Assessment: Our Approach*, London.

DEFRA (2010), *UK Food Security Assessment: Detailed Analysis*, London.

Dodge, E. and S. Gemessa. (2012) "Food Security and Rice Price Stabilization in Indonesia: Analysis of Policy Responses", second year paper of the MPA/ID Program, Harvard Kennedy School of Government.

Drèze, J. and A. Sen (1990), *The Political Economy of Hunger*, Calderon Press, Oxford.

Drichoutis, A.C. et al. (2008), "Household food consumption in Turkey: a comment", *European Review of Agricultural Economics*, Vol.35(1), pp. 93-8.

Elci, C. (2006), "The Impact of HPAI of the H5N1 Strain on Economies of Affected Countries", London South Bank University, London.

Fane, G. and P. Warr (2008), "Agricultural Protection in Indonesia", *Bulletin of Indonesian Economic Studies*, Vol. 44, pp. 133-50.

Fane, G. and P. Warr (2009). "Indonesia" in Kym Anderson and Will Martin (eds), *Distortions to Agricultural Incentives in Asia*, World Bank, Washington DC, pp. 165-196.

FAO (2013), *The State of Food Insecurity in the World 2013*, Food and Agriculture Organization of the United Nations Publications, Rome. http://www.fao.org/publications/sofi/2013/en/.

FAO (2008), "FAO methodology for measurement of food deprivation". Food and Agriculture Organization of the United Nations Statistics Division, Rome.

FAO (2006), "Food Security", *Briefing Note*, Issue 2, May, Food and Agriculture Organization of the United Nations Publications, Rome.

FAO (2003), "Food Security: Concepts and measurement", Chapter 2 in FAO (2003), *Trade reforms and Food security: Conceptualizing the Linkages*, Food and Agriculture Organization of the United Nations Publications, Rome.

FAO (2001), *The State of Food Insecurity*, Food and Agriculture Organization of the United Nations Publications, Rome.

FAO, IFAD, IMF, OECD, UNCTAD, WFP, the World Bank, the WTO, IFPRI and the UN HLTF (2011), "Price Volatility in Food and Agricultural Markets: Policy Responses", Report for the G20.

Flordeliza A. et al. (2013), "Estimating the Demand Elasticities of Rice in the Philippines", Southeast Asian Regional Center for Graduate Study and Research in Agriculture, Los Baños, Laguna, Philippines.

Gebhart, J.C. (1920) "Poverty and Malnutrition", *The Public Health Journal*, Vol. 11, No. 8, pp. 371-379.

Gignoux J. and M. Menéndez (2013), "Short and Long Run Effects of Earthquakes on Household Welfare in Indonesia", http://www.iza.org/conference_files/worldb2013/menendez_m9061.pdf.

Green, R. and J.M. Alston (1990), "Elasticities in AIDS models", *American Journal of Agricultural Economics*, Vol. 72, No. 2, pp. 442-445.

Hall, D.H., C. Benigno, and W. Kalpravidh (2006). "The Impact of Avian Influenza on Small and Medium Scale Poultry Producers in South East Asia (preliminary findings)", selected paper prepared for presentation at the American Agricultural Economics Association Annual Meeting, Long Beach, California, 23-26 July.

Heong, K.L. and K.G. Schoenly (1998), "Impact of Insecticides on Herbivore-Natural Enemy Communities in Tropical Rice Ecosystems", in P.T. Haskell and P. McEwen (Eds.), *Ecotoxicology: Pesticides and Beneficial Organisms*, Chapman and Hall, London.

Heong, K., L. Wong and J.H. Delos Reyes (2013), "Addressing Planthopper Pest Outbreak Threats to the Sustainable Development of Asian Rice Farming and Food Security: Fixing the Insecticide Misuse," *ADB Sustainable Development Working Paper*, No. 27, Asian Development Bank, Manila, Philippines.

Hill, H. (2012). *The Indonesia Economy*, Cambridge University Press, Cambridge.

IFPRI (2013). *Global Hunger Index*. http://www.ifpri.org/publication/2013-global-hunger-index.

Jayne, T.S. and S. Rashid (2013), "Input Subsidy Programs in Sub-Saharan Africa: A Synthesis of Recent Evidence", *Agricultural Economics* Vol. 44, pp. 547–562.

Johnson C. (1998), "Survey of Recent Developments", *Bulletin of Indonesian Economic Studies*, Vol. 34:2, pp. 3-59, http://dx.doi.org/10.1080/00074919812331337320.

Markowitz, H.M. (1952), "Portfolio Selection", *The Journal of Finance*, Vol. 7, No. 1, pp. 77–91.

Masset, E. (2011), "A Review of Hunger Indices and Methods to Monitor Country Commitment to Fighting Hunger", *Food Policy*, Vol. 36, pp. S102-S108.

MDB (2011), "Multilateral Development Banks' State of Play on Risk management Tools and Instruments", document prepared for the G20 summit, 17 September.

Naylor, R.L. et al. (2001), "Using *El Niño*/Southern Oscillation Climate Data to Predict Rice Production in Indonesia", *Climate Change,* Vol 50, pp. 255-265.

OECD (2015), *OECD Economic Surveys: Indonesia 2015*, OECD Publishing, Paris. DOI: http://dx.doi.org/10.1787/eco_surveys-idn-2015-en.

OECD (2014), "Measuring the Incidence of Policies along the Food Chain", OECD document [TAD/CA/APM/WP(2013)30/FINAL].

OECD (2013a), *Global Food Security: Challenges for the Food and Agricultural System*, OECD Publishing, Paris.
DOI: http://dx.doi.org/10.1787/9789264195363-en.

OECD (2013b), "Smallholder Risk Management in Developing Countries", *OECD Food, Agriculture and Fisheries Papers*, No. 61, OECD Publishing, Paris.
DOI: http://dx.doi.org/10.1787/5k452k28wljl-en.

OECD (2013c), *Agricultural Policy Monitoring and Evaluation 2013: OECD Countries and Emerging Economies*, OECD Publishing, Paris.
DOI: http://dx.doi.org/10.1787/agr_pol-2013-en.

OECD (2011), *Managing Risk in Agriculture: Policy Assessment and Design*, OECD Publishing, Paris.
DOI: http://dx.doi.org/10.1787/9789264116146-en.

OECD (2009), *Managing Risk in Agriculture: A Holistic Approach*, OECD Publishing, Paris.
DOI: http://dx.doi.org/10.1787/9789264075313-en.

OECD (2012a), *Agricultural Policies for Poverty Reduction*, OECD Publishing, Paris.
DOI: http://dx.doi.org/10.1787/9789264112902-en.

OECD (2012b), *Review of Agricultural Policies: Indonesia 2012*, OECD publishing, Paris, http://dx.doi.org/10.1787/9789264179011-en.

OECD (2006), *Agricultural and Fisheries Policies in Mexico: Recent Achievements, Continuing the Reform Agenda*, OECD Publishing, Paris.
DOI: http://dx.doi.org/10.1787/9789264030251-en.

OECD (2001), *Market Effects of Crop Support Measures*, OECD Publishing, Paris.
DOI: http://dx.doi.org/10.1787/9789264195011-en.

OFDA/CRED, EM-DAT, The OFDA/CRED International Database, http://www.grida.no/geo/GEO/Geo-2-330.htm.

Pardede, R. (1999) "Survey of Recent Developments", *Bulletin of Indonesian Economic Studies,* Vol. 35/2, August, pp.3-39.

Pingali, P.L., M.H. Hossain, and R. Gerpacio (1997), "Asian Rice Bowls: The Returning Crisis?" CAB International, Oxford, in association with the International Rice Research Institute and CABI International, Manilla.

Radhakrishna, R. et al. (2004), "Chronic Poverty and Malnutrition in 1990s", *Economic and Political Weekly*, Vol. 39, No. 28, pp. 3121-3130.

Ravallion, M. and B. Bidani (1994), "How Robust is a Poverty Profile?", *The World Bank Economic Review*, Vol. 8, N° 1, pages 75-102.

Salafsky, N. (1994) "Drought in the Rain Forest: Effects of the 1991 *El Niño*-Southern Oscillation Event on a Rural Economy in West Kalimantan, Indonesia", *Climate Change,* Vol. 27, pp. 373-396.

Schmidhuber, J. and F.N. Tobiello (2007), "Global Food Security Under Climate Change", *Proceedings of the National Academy of Sciences*, N°104 (50), pp. 19703-19708.

Sen, A. (1981), *Poverty and Famines: An essay on entitlements and Deprivation*, Oxford University Press, Oxford.

Shonkwiler, J.S. and S.T. Yen, 1999, "Two-Step Estimation of a Censored System of Equations", *American Journal of Agricultural Economics*, Vol. 81, pp.972-82.

Sibrián, R. (2009), "Indicators for Monitoring Hunger at Global and Subnational Levels", *FAO Statistics Division Working Papers Series*, No. ESS/ESSG/013e, April.

Tafere K., A. Seyoum Taffesse, and S. Tamiru S., with Tefera, S. and Paulos, Z. (2010), "Food Demand Elasticities in Ethiopia: Estimates Using Household Income Consumption Expenditure (HICE) Survey Data", Ethiopia Strategy Support Program 2 (ESSP2), International Food Policy Research Instituten, Washington DC and Ethiopian Development Research Institute, Addis Ababa.

Timmer, P. (2008), "Causes of High Food Prices", *ADB Economic Paper Series* No. 128, Asian Development Bank, Manila.

Townsend, R.M. (2013), "Accounting for the Poor", *American Journal of Agricultural Economics*, Vol. 95, No. 5, pp. 1196-1208.

Thomas, D. and E. Frankenberg (2007), "Household Responses to the Financial Crisis in Indonesia: Longditudinal Evidence on Poverty, Resources and Well-being", in A. Harrison (ed.), *Globalization and Poverty*, University of Chicago Press, Chicago.

United Nations (2014), Millennium Development Goals Report, UN, New York, http://www.un.org/millenniumgoals/2014%20MDG%20report/MDG%202014%20English%20web.pdf.

Warr. P. (2011). "Petroleum Prices and Poverty in a Transport-poor economy", *ASEAN Economic Bulletin* Vol. 28, pp. 257-280.

Warr, P. and A.A. Yusuf (2014). "World Food Prices and Poverty in Indonesia". *Australian Journal of Agriculture and Resource Economics*, Vol. 57, pp 1-21.

Wittwer, G. (2012), "Practical Policy Analysis Using TERM", Chapter 1 in G. Wittwer (Ed.), *Economic Modeling of Water: The Australian CGE Experience*, Springer, Dordrecht, Netherlands.

Wooldridge, J.M. (2002), *Econometric Analysis of Cross Section and Panel Data*, MIT press, Boston.

World Bank (2012a), "Raskin Subsidized Rice Delivery", *Public expenditure review (PER) Social Assistance Program and Public Expenditure Review*, No. 3. Washington, DC. http://documents.worldbank.org/curated/en/2012/02/15893857/raskin-subsidized-rice-delivery.

World Bank (2012b), *Bantuan Langsung Tunai (BLT) Temporary unconditional cash transfer. Public expenditure review (PER); Social assistance program and public expenditure*, Review No. 2. Washington, DC. http://documents.worldbank.org/curated/en/2012/02/15893823/bantuan-langsung-tunai-blt-temporary-unconditional-cash-transfer.

World Bank (2004), *Beyond Economic Growth: An Introduction to Sustainable development*, WBI Learning Resource series, Washington DC. http://www.worldbank.org/depweb/english/beyond/beyondco/beg_06.pdf.

World Bank (1986), *Poverty and Hunger: Issues and Options for Food Security in Developing Countries,* A World Bank Policy Study, Washington DC. http://documents.worldbank.org/curated/en/1986/07/440681/poverty-hunger-issues-options-food-security-developing-countries.

World Food Programme (2010), *Indonesia - Food Security and Vulnerability Atlas, March 2010.* UN World Food Programme, http://www.wfp.org/content/indonesia-food-security-and-vulnerability-atlas-2009.

Yen, S.T., B. Lin, and D.M. Smallwood (2003), "Quasi- and Simulated-Likelihood Approached to Censored Demand Systems: Food Consumption by Food Stamp Recipients in the United States", *American Journal of Agricultural Economics*, Vol.85 (May), pp.458-78.

Yusuf, A.A. (2008), "INDONESIA-E3: An Indonesian Applied General Equilibrium Model for Analyzing the Economy, Equity, and the Environment", *Working Papers in Economics and Development Studies (WoPEDS),* No 200804, Department of Economics, Padjadjaran University, http://lp3e.fe.unpad.ac.id/wopeds/200804.pdf.

Yusuf, A.A. (2006). "Constructing Indonesian Social Accounting Matrix for Distributional Analysis in the CGE Modelling Framework", *Working Papers in Economics and Development Studies,* No. 200604, Department of Economics, Padjadjaran University, Bandung, Indonesia.

# ORGANISATION FOR ECONOMIC CO-OPERATION AND DEVELOPMENT

The OECD is a unique forum where governments work together to address the economic, social and environmental challenges of globalisation. The OECD is also at the forefront of efforts to understand and to help governments respond to new developments and concerns, such as corporate governance, the information economy and the challenges of an ageing population. The Organisation provides a setting where governments can compare policy experiences, seek answers to common problems, identify good practice and work to co-ordinate domestic and international policies.

The OECD member countries are: Australia, Austria, Belgium, Canada, Chile, the Czech Republic, Denmark, Estonia, Finland, France, Germany, Greece, Hungary, Iceland, Ireland, Israel, Italy, Japan, Korea, Luxembourg, Mexico, the Netherlands, New Zealand, Norway, Poland, Portugal, the Slovak Republic, Slovenia, Spain, Sweden, Switzerland, Turkey, the United Kingdom and the United States. The European Union takes part in the work of the OECD.

OECD Publishing disseminates widely the results of the Organisation's statistics gathering and research on economic, social and environmental issues, as well as the conventions, guidelines and standards agreed by its members.

OECD PUBLISHING, 2, rue André-Pascal, 75775 PARIS CEDEX 16
(51 2015 05 1) ISBN 978-92-64-23386-7 – 2015

www.ingramcontent.com/pod-product-compliance
Lightning Source LLC
LaVergne TN
LVHW081416110826
845149LV00010B/1760

* 9 7 8 9 2 6 4 2 3 3 8 6 7 *